Nanotechnology: A Very Short Introduction

VERY SHORT INTRODUCTIONS are for anyone wanting a stimulating and accessible way into a new subject. They are written by experts, and have been translated into more than 45 different languages.

The series began in 1995, and now covers a wide variety of topics in every discipline. The VSI library currently contains over 700 volumes—a Very Short Introduction to everything from Psychology and Philosophy of Science to American History and Relativity—and continues to grow in every subject area.

Very Short Introductions available now:

Available soon:

ANCIENT GREEK AND ROMAN
   SCIENCE Liba Taub
ADDICTION Keith Humphreys
VATICAN II Shaun Blanchard &
   Stephen Bullivant

OBSERVATIONAL ASTRONOMY
   Geoff Cottrell
MATHEMATICAL ANALYSIS
   Richard Earl

For more information visit our website

www.oup.com/vsi/

Philip Moriarty

NANOTECHNOLOGY

A Very Short Introduction

OXFORD
UNIVERSITY PRESS

OXFORD
UNIVERSITY PRESS

Great Clarendon Street, Oxford, OX2 6DP,
United Kingdom

Oxford University Press is a department of the University of Oxford.
It furthers the University's objective of excellence in research, scholarship,
and education by publishing worldwide. Oxford is a registered trade mark of
Oxford University Press in the UK and in certain other countries

Published in the United States of America by Oxford University Press
198 Madison Avenue, New York, NY 10016, United States of America

British Library Cataloguing in Publication Data

Data available

Library of Congress Control Number: 2022941261

ISBN 978-0-19-884110-4

Printed and bound by
CPI Group (UK) Ltd, Croydon, CR0 4YY

Contents

Preface

'nano, /ˈnanəʊ/. From the Greek verb meaning "to attract research funding".'

I'm a self-confessed nanoscientist but have a soft spot for the tongue-in-cheek definition of *nano*—due to George Smith, emeritus Professor of Materials at the University of Oxford—given above. Smith's cynicism highlights a key issue that looms large over any introduction to the science and techology of the ultrasmall: those at the more sceptical end of the spectrum of opinion would argue that nanotechnology is nothing more than advanced chemistry, condensed matter physics, and/or materials science in action, and that the now ubiquitous 'nano' prefix is more a canny marketing tool than a scientific descriptor. As Kevin Kelleher perceptively put it in 'Here's why nobody's talking about nanotech anymore', published in *Time* magazine in October 2015:

> Of all the investment fads and manias over the past few decades, none have been as big of a fizzle as the craze for nanotech stocks. Ten years ago, venture capitalists were scrambling for investments, start-ups with 'nano' in their names flourished and even a few nanotech funds launched hoping to track a rising industry.
>
> And today? Nobody in the stock market gets excited about the phrase 'nanotech' anymore.

Even nanoscientists themselves are jaded by the nanohype. Shortly after I was asked to write this book, a scientific paper on the *wunderkind* material graphene (see Chapter 3) was published in the prestigious American Chemical Society journal *ACS Nano*, with the brilliantly biting title of 'Will any crap we put into graphene increase its electrocatalytic effect?' (L. Wang, Z. Sofer, and M. Pumera, *ACS Nano* **14**, 21 (2020)).

Ever since its inception back in the late 1970s/early 1980s, there has been too much of a willingness in the nanoscience community to make overblown promises about the real world applicability of both fundamental and entirely mundane discoveries in the field. In parallel, the 'nano' prefix has been saddled to an impressively wide variety of nouns and verbs, often in a bid to give a paper or grant proposal a suitably *en vogue* sheen: nanoparticles, nanotubes, nanoclusters, nanobubbles, nanochips, nanocubes, nanoguitars, nanohacking, nanosculpting, nanomedicine, nanoelectronics, nanobiotech, nanopants, and even a family of nanovegetables (nanocabbage, nanobroccoli, nanocauliflower, and nanocarrots—I kid you not) have each been coined and used in scientific and/or *pop sci* accounts of advances in nanotech. And that's certainly not an exhaustive list. (One might say that it is but a nanoscopic sampling.)

In *Nanohype* (Prometheus Books, 2006), David Berube quotes US Senator Ron Wyden speaking during the hearings for the 21st Century Nanotechology Research & Development Act (which was signed into law by President George W. Bush in December 2003),

> The joke these days in the world of science is that everyone is doing nano work. Just as the nineties saw everyone putting dot.com after titles, everyone is putting nano before their science.

So even two decades ago (at the time of writing) we'd apparently heard more than enough of the nano buzzword. One might then quite reasonably ask whether nanotechnology is nothing more than a case of smoke and microscopic mirrors. Is it really just a matter of hype, hyperbole, and hubris? Or is there something more substantial in the invisible? Could those billions of pounds/dollars/ euros (insert currency of choice) of investment by governments worldwide, those Nano Research™ institutes that sprang up virtually overnight, and the exponential growth in scientific journals with 'nano' somewhere in their title all be a hoax?

This *Very Short Introduction* presents the case for nanotechology in the face of all the cynicism. Whatever criticisms might justifiably be made about the overuse of that prefix, science and technology at the nanoscale—and we'll cover in some detail just what that means in practice—are nonetheless exceptionally exciting, groundbreaking, and, at best, revolutionary areas of research and R&D. So much of the world around us, including essentially all of 21st-century computing technology, involves science and engineering at the nanometre scale.

From the fundamental to the practical, from the most esoteric quantum mechanics to the bounciness of tennis balls, nanotechnology now underpins how we probe and control matter in ways that would have been unfathomable even a generation ago. *A Very Short Introduction to Nanotechnology* is my attempt to capture the essence of the nanoscale; to explain just why nanoscience is different from—but also an amalgam of—traditional scientific disciplines, and to highlight some of the most fascinating research areas in the field. These range from the computer-controlled manipulation of matter at the single chemical bond limit to the development of nanomachines that owe their operation to nanoscientists' harnessing of nature's design principles.

In a field that moves as quickly as nanotechnology, I can only provide a snapshot of the state of the art at the time of writing.

I've aimed to make the examples I've chosen to illustrate the science as topical as possible but suspect that if a new edition of this *Very Short Introduction* is required, it will be sooner rather than later.

As an avid reader of the *Very Short Introduction* series, I was delighted to receive an invitation from Latha Menon, Senior Commissioning Editor at Oxford University Press, to write a VSI on nanotechnology. That was four years ago. My sincere thanks to Latha and her colleague Jade Dixon, Project Editor at OUP, for their immense patience with my missing of deadlines throughout that time. Latha's edits and feedback have been indispensable throughout, both with regard to finding the correct tone for the VSI series and in excising my voluble off-topic rambling. A VSI needs to get to the point as quickly as possible—hardly my forte—and Latha and Jade's edits and suggestions were invaluable in cutting my writing down to size. I'm also very grateful to the anonymous reviewers of both the book proposal and the final draft of the manuscript for their very helpful comments and advice.

Philip Moriarty
Nottingham, March 2022

List of illustrations

List of illustrations

Chapter 1
Welcome to NanoPut

The NanoPutians are a family of anthropomorphic molecules designed and created by Stephanie H. Chanteau and James M. Tour of the Centre for Nanoscale Science and Technology at Rice University. They form a remarkable demonstration of the exquisite control of the atomic and molecular building blocks of our universe that is now possible. One member of that family, NanoKid, is sketched in Figure 1. Standing only two nanometres tall, he's almost exactly 100,000,000 times smaller than the six-inch Lilliputians in Jonathan Swift's timeless *Gulliver's Travels*. Or, to put that in a rather more immediate context, the full stop at the end of this sentence is roughly 150,000 NanoKids wide. Not just microscopic but *nanoscopic*, each of NanoKid's atoms has been put in place via synthetic reactions chosen on the basis of what is best described as a molecular blueprint; billions upon billions of identical NanoPutians emerge from the chemical soup.

Equally impressively, the top and bottom halves of NanoKid's molecular skeleton were synthesized separately and then 'stitched' together at the waist; he's a nanoscale Frankenstein's monster. That's an astonishing feat of chemical synthesis, exploiting an approach that some have dubbed molecular surgery. And yet there are many critics of nanotech who would claim that, impressive though NanoKid and his siblings might be, there's little that's

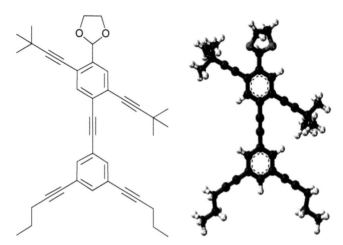

1. 'NanoKid', one of a family of NanoPutian molecules synthesized by Stephanie H. Chanteau and James M. Tour of the Center for Nanoscale Science and Technology at Rice University. A chemical bonding model is shown on the left and the corresponding molecular structure is on the right, in which carbon, hydrogen, and oxygen atoms are shown as black, white, and grey spheres, respectively.

really *new* here. Isn't nanotechnology really just chemistry in disguise? Isn't it all just a case of the Emperor's New Clothes writ small? Doesn't the hype outweigh the hard science?

My aim in this *Very Short Introduction* is to cut through the hyperbole, focus on the fundamental scientific principles at the core of nanoscience, and show that, despite the (sometimes valid) criticisms of the naysayers, nanotechnology *is* new, exciting, and different. Not only can we image and precisely position individual atoms, but also the state of the art in nanotech involves computer-directed targeting of single chemical bonds within an individual molecule; that's a level of control that even just a few decades ago seemed an impossible goal. Moreover, these scientific advances are increasingly transferred to technological advances (and vice versa) in an ever-growing list of products spanning everything from sunscreen to laptops, water purifiers to golf balls, and solar cells to smart fabrics that never stain. As we're about to see, developments in nanotechnology sometimes deserve the hype they attract.

Before we start our whistle-stop tour of the nanoworld, however, we need to define just what we mean by the terms *nanotechnology* and *nanoscience*. This would seem an entirely appropriate place to begin, but establishing a universally agreed definition of nanotechnology turns out to be remarkably difficult. The *nano* prefix refers to a nanometre (nm), 10^{-9} m. That's half a NanoKid in length, or, to use a rather more general comparative measure, if the diameter of a marble were one nanometre then the Earth would be approximately a metre in diameter. As another helpful nano-comparison, consider that a sheet of paper is roughly 100,000 nanometres thick. Nanoscience, however, involves the study of structures and devices that can be considerably greater than 1 nm across. How then do we choose the upper (and, indeed, lower) size limits to distinguish nanoscience and nanotechnology from, say, microtechnology? Some have suggested that the upper 'cut-off' for nanotechnology is 100 nm. This, however, is also fraught with difficulties because one could ask 'why not 500 nm? Or, indeed, why not 101 nm? Or 99 nm?'

Instead of fruitlessly and pedantically trying to define nanoscience in terms of particular length scales, the Royal Society, London, put forward the following definition in an influential report in 2004:

> Nanoscience is the study of phenomena and manipulation of materials at atomic, molecular and macromolecular scales, *where properties differ significantly from those at a larger scale.* Nanotechnologies are the design, characterisation, production and application of structures, devices and systems by controlling shape and size at nanometre scale.

I have highlighted the key phrase, which, although still somewhat vague, cuts to the crux of what is special about nanoscience and nanotechnology. Simply by changing the size of an object at the atomic, molecular, or, as the Royal Society puts it in the definition above, macromolecular scale, one can radically change its properties. As we'll see later, the size-dependent change can be particularly striking when the size of a nanoparticle becomes comparable to the wavelength of its electrons, that is when quantum physics plays a key role.

Although it is therefore best to avoid definitions of nanoscience in terms of stringent length scale limits, a good 'rule of thumb' is that if the structure or process in which you are interested can't be studied using conventional optical microscopy then it is nanoscopic rather than microscopic. We are going to see soon that the field of nanotechnology owes much, if not all, of its origins to the invention of an entirely new breed of ultrahigh resolution microscope, capable of resolving individual atoms and molecules but involving no optics at all (i.e. no lenses, mirrors, etc.).

True innovation stems from crossing traditional boundaries and this is where nanotech and nanoscience excel: at the nanometre level, the rather artificial (and largely historical) divisions between physics, chemistry, biology, and materials science become blurred. When we are probing the electronic structure of a single atom, or manipulating single chemical bonds within a molecule, is what we're doing physics, chemistry, or materials science? What if we're measuring the mechanical strength of a protein when we apply a force to one part of the molecule, or stretching a DNA strand and monitoring its response? Is that biology, physics, or chemistry? Or is it instead cross-disciplinary research, representing a refreshing and, at best, revolutionary mix of the traditional sciences, from which stem insights that are simply not possible if we're stuck within a particular disciplinary silo?

Interdisciplinarity is the bedrock of nanotechnology, but each nanoscientist brings their own disciplinary baggage to the field. Our nanoscience research group at the University of Nottingham, for example, is almost entirely made up of physicists and yet much (if not all) of the research we do could equally well be described as physical chemistry, surface science, or materials science. That physics 'upbringing', however, means that my emphasis throughout this book will be skewed slightly more towards the physical science aspects of nanotech, although I'll certainly not be eschewing the life sciences.

With that particular bias declared, let's begin this whistle-stop tour of the nanoscale by introducing the remarkable instrument to which I alluded above, namely the scanning probe microscope (SPM). Probe microscopes are central to our ability to map and manipulate matter at the atomic and (sub)molecular levels. Indeed, many nanoscientists, regardless of their disciplinary background, would claim with some justification that Year Zero for nanotech was 1981, due to the invention that year of the probe microscope at IBM Research Labs in Ruschlikon, Zurich. SPMs allow us to not only measure forces between atoms and molecules but also to control those interatomic and intermolecular interactions so as to move matter with atomic (or even better-than-atomic) precision.

With a state-of-the-art probe microscope we can push, pull, prod, poke, and pick up atoms one at a time, under computer control, and with an accuracy that is well outside the capabilities of any other scientific instrument. But what does it mean to push or pull an atom? What forces are involved? How much energy does it take? And how do we apply and target forces on such a small scale?

Feeling the force

Conceptually, the probe microscope could, from a certain perspective, be said to be much less complicated than its

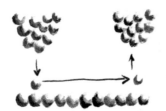

2. The scanning probe microscope achieves atomically precise imaging and manipulation of matter by scanning an ultrasharp tip, terminated by a single atom, very close to a surface.

traditional optical counterpart. Instead of bending light waves in just the right way to form a magnified image, a probe microscope allows us to both see and move individual atoms by scanning an atomically sharp tip back and forth across a surface (see Figure 2). Initially, the tip is a sharpened metal wire (usually tungsten, although both gold and an alloy of platinum and iridium are also commonly used). However, the apex of the tip will often be deliberately or inadvertently terminated by an atom from the sample surface; when this process is carefully controlled it is known as tip functionalization and provides a high degree of control of the atomistic structure of the probe.

The image on the right in Figure 2 is the first demonstration of writing with atoms carried out in 1990 by Don Eigler and Eric Schweizer of IBM Research Labs. Each bright feature is a single Xe atom (on an ultracold nickel surface) moved in place using the 'sliding' strategy shown in the cartoon. The arrows in that cartoon show how an individual atom is positioned by changing the tip-sample separation, thus controlling the force between the atom selected for manipulation and the probe. The tip is first lowered—increasing the interaction of the chosen atom—and then moved a fixed distance parallel to the surface, taking the atom with it. At the end of the manipulation step, the tip is retracted back to its initial (imaging) height. By keeping the distance between the probe and the surface very small—generally less than a nanometre, that is a few atomic diameters or less—the interaction between the atoms at the apex of the probe and those at the surface of the sample below can be mapped out atom by atom. In fact, the resolution can be much, much better than a single atom; as we'll see soon, the state of the art involves mapping out the chemical architecture, that is the bonding, *inside* individual molecules.

The form of the probe-sample interaction depends on just what we want to map and measure. There is now a growing family of probe microscopes, each tuned to exploit one or more of the forces and interactions that exist between atoms, molecules, and/or nanoparticles (i.e. tiny chunks of matter comprising anywhere between a few to a few thousand atoms—more on these later).

Physicists classify forces into four fundamental classes: gravity, the strong and weak nuclear forces, and electromagnetism. When it comes to the nanoscale, however, electromagnetic interactions dominate by quite some margin. We can do a simple back-of-the-envelope calculation to highlight just how little influence gravity, for one, has on the interaction between two atoms or molecules. A simple equation known as Coulomb's law

$$F_{el} = kq^2 / r^2$$

gives us the magnitude of the electrostatic force, F_{el}, between two equal charges, q, separated by a distance r. (Here, k is a constant.) Let's consider two atoms, one which has lost an electron, and another which has gained an electron, so they both have a net charge and therefore are *ions*. A good example is the case of Na^+ and Cl^-, a net positively charged sodium ion and a chlorine ion with a net negative charge, due to the loss or addition of an electron, respectively. We'll assume that the Na^+ and Cl^- are separated by the same distance as they are in the NaCl (i.e. salt) lattice: 0.236 nm.

The electrostatic force between Na^+ and Cl^- is easily calculated by plugging the appropriate values into the equation above. If we do this we find that the force is about 4 nanonewtons (nN). This is a tiny value as compared to, for example, the weight of a typical adult (which is of the order of hundreds of newtons). But that nanoscale force is beyond vast if we compare it to the gravitational force between the atoms.

To calculate the corresponding gravitational force, F_g, between Na$^+$ and Cl$^-$, we use Newton's law of gravitation, which is of a very similar form to that of the electrostatic force,

$$F_g = Gm_{Na}m_{Cl} / r^2,$$

where r is once again the separation, m_{Na} and m_{Cl} are the mass of the Na and Cl ions, respectively, and G is a number known as Newton's gravitational constant. The gravitational force between the atoms is of order 10^{-42} N—inconceivably tiny. If we take the ratio of the electrostatic and gravitational forces between Na$^+$ and Cl$^-$, the r^2 in each cancel out and we obtain a factor of 10^{33}. This is not just a large number, it's staggeringly, mind-blowingly *huge* (because atoms have very little mass indeed). The only role that gravity typically plays in nanoscience experiments is to keep the equipment—and the scientists—firmly anchored to the ground (because those are massive objects, comprising countless atoms and molecules).

Similarly, when it comes to the strong and weak fundamental forces, which hold sway between the protons and neutrons of the atomic nucleus, the forces and energy scales that we explore and exploit in nanotechnology are nowhere near the nuclear limit. Before comparing the nano and nuclear scales in detail, we need to consider the most appropriate units to use for our measure of energy. Although the standard (SI) unit for energy is the joule (J), on the nanometre, and (sub)nuclear, scale this is much too coarse to be of much use. Instead, we use the electron-volt (eV) as the energy unit: 1 eV is the energy that an electron acquires if accelerated through a potential difference of 1 V. In the nanotech context, however, much more important than the textbook definition of the electron-volt is that a typical chemical bond has an energy of a few eV: it takes an injection of energy of this order to split up a pair of atoms. This contrasts dramatically with the hundreds of millions of electron-volts, i.e. ~100 MeV, characteristic of the binding energy of the constituents of an atomic nucleus.

This means that nanotechnology is rooted entirely in the study and manipulation of electromagnetic interactions (including electrostatic forces) between atoms and molecules; neither gravity nor the strong or weak nuclear force need concern us. Instead, it is the electromagnetic forces between *electrons* that are our focus. But then so very much of the world around us, including every electronic or electrical device—every smartphone, every laptop, every kitchen appliance—is fundamentally defined by the interactions of electrons.

Why is diamond more rigid than putty? Because of the relative strength of its chemical bonds, that is the electron–electron interactions. Why is glass transparent while gold is golden? Because of the interaction (or relative lack thereof) of light, an electromagnetic wave, with the *electrons* in each case. Why is iron magnetic but aluminium apparently oblivious to the influence of a nearby magnet? Because of the quantum mechanical properties of the electrons. And why does copper wire conduct electricity while its surrounding plastic insulation carries no current? Again, the arrangement of the electrons in each material holds the answer.

The interactions of electrons underpin the material world, including all information technology, and a very large amount of nanotechnology and nanoscience is therefore especially concerned with measuring and modifying electronic behaviour. By controlling and corralling electrons we can *tune* the properties of a material: change its colour, strength, ability to conduct electricity, its chemical reactivity, and its reaction to a range of stimuli including heat, light, and strain. Nanotechnology provides the tools to design and realize bespoke materials whose nanoscale, atomic, and/or molecular structure has been controlled from the bottom up: atom by atom. Gaining this level of precision involves the control of interatomic forces, and a deep understanding of just how those forces depend on the separation of the atoms.

At this juncture I'm going to turn to a particularly apposite quote from Richard Feynman, the (in)famous 20th-century physicist, raconteur, and player of bongo drums. In the context of nanotechnology, Feynman is almost always cited because of a prescient after-dinner talk he gave to the American Physical Society back in 1959 entitled 'There's plenty of room at the bottom', which remarkably predicted the type of single atom manipulation that is now not only possible but also increasingly *de rigueur* in much of nanoscience and nanotechnology. It's become something of a cliché to quote at length from this talk when describing nanotech so I'm going to forgo that particular reference for now. Instead, I'm going to turn to Feynman's famed *Lectures in Physics*, in which he describes the atomistic structure of matter and the nature of interatomic interactions as the most important piece of scientific information that humanity has discovered:

> If, in some cataclysm, all of scientific knowledge were to be destroyed, and only one sentence passed on to the next generations of creatures, what statement would contain the most information in the fewest words? I believe it is the atomic hypothesis (or the atomic fact, or whatever you wish to call it) that all things are made of atoms—little particles that move around in perpetual motion, attracting each other when they are a little distance apart, but repelling upon being squeezed into one another.

Why is it that atoms, as Feynman puts it, attract each other when they are 'a little distance apart' and yet repel 'upon being squeezed into one another'? The answer to this question is central to nanotechnology.

A close bond

The central graph of Figure 3 shows how the force between two xenon (Xe) atoms varies as their separation, r, changes. I've chosen Xe due to its particularly important place in the history of nanotechnology—it was the first atom to be precisely positioned under computer control (see Figure 2). At large values of r, the Xe atoms interact very weakly; the force (and potential energy) associated with their interaction is extremely small, ever-approaching zero as the interatomic separation increases. If, instead, the atoms are brought closer together they feel an attractive interaction, the force becoming increasingly more negative until it reaches a minimum. (A negative force is associated with an attraction, whereas repulsion gives rise to a positive force.) As the atoms are moved still closer, the force starts to increase until it again reaches 0—at this point the atoms are at their *equilibrium separation*. They're close enough to be interacting but there is no net force; the forces on the atoms balance out. (I'll explain the origin of those opposing forces very soon.) The equilibrium separation defines the *bond length* for an interacting pair of atoms or molecules.

If we now try to push the atoms together beyond this point, that is to reduce their separation below the equilibrium value, the force becomes positive and increases extremely rapidly with even small changes in interatomic separation. There is a *very* strong repulsive interaction—not just sub-nanometre, but sub-Ångstrom, displacements of the atoms towards each other cause the force (and energy) to rise dramatically. (An Ångstrom is 0.1 nm. It's a unit used extensively by scientists interested in the atomic structure of materials because the lengths of chemical bonds are typically of the order of a few Ångstroms.) This combination of an attractive and a repulsive interaction gives rise to the characteristic shape of the force-vs-separation curve shown in Figure 3. If we want to break the chemical bond then we need to

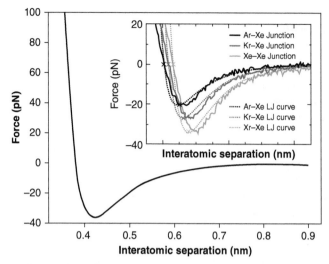

Nanotechnology

3. The variation in force between two atoms as their separation is varied, calculated using the Lennard-Jones potential described in the text. The inset shows equivalent experimental measurements, labelled 'Junction', and their comparison to the appropriate Lennard-Jones (LJ) potential in each case, of the interatomic force for three different pairs of inert gas atoms: xenon–xenon, krypton–xenon, and argon–xenon.

supply enough energy—in the form of, for example, heat, light (i.e. electromagnetic radiation), or, as we shall soon see, even via a highly targeted mechanical force—so as to allow the atoms to overcome the mutual interaction and escape each other.

Figure 3 was calculated using a very simple empirical formula for interatomic interactions known as the Lennard-Jones potential, named after the mathematician and theoretical physicist John Edward Lennard-Jones. The Lennard-Jones potential is very often used to account for the interactions of inert atoms such as those of xenon, a noble gas. You might reasonably ask why two xenon atoms interact at all, given that they are inert. Noble gases are, after all, often chosen for applications where a lack of chemical reactivity is a must because they have complete shells of electrons. Moreover, there's no net charge—we're considering neutral atoms (rather than ions like Na⁺ and Cl⁻). Yet the xenon atoms still attract each other, just as Feynman described. Why?

An atom is generally visualized as a miniature solar system with the nucleus at the centre and the electrons circling in well-defined orbits. This is the Bohr model, named after the Danish physicist Niels Bohr, and is a picture of the atom that is now a little over a century old. Although it's not entirely unhelpful—many scientists still often conceptualize the atom in this way—the Bohr model is very badly wrong. Electrons do not orbit like tiny planets around a star. Quantum mechanics tells us that they are best described by probability clouds whose overall shape depends on the energy of the electron. We'll have much more to say about the role of quantum mechanics in nanotechnology in the next chapter but for now the core aspect of quantum mechanics that is relevant to interatomic forces is that *fluctuation* is everything at the nanoscale. The probability cloud that describes the electrons' positions means that they can be found at very different places across the atom, quite unlike the restricted, regular orbits of the Bohr model.

The fluctuations of the electrons in turn mean that at any given moment of time there is a charge imbalance across the atom, giving rise to an electric dipole: a separation of net positive and net negative charge. Where there is a local increase in the probability of finding an electron, the atom is slightly more negatively charged, and where the fluctuations are such that there is a lower chance of finding an electron, there will be a net positive region. A dipole on one atom (or molecule) induces a dipole of the opposite sense on another, leading to an attraction. It is this effect, known as the London dispersion force—after the physicist Fritz London, who explained the origin of the attractive interaction back in 1930—that is responsible for even neutral atoms attracting each other. (I should note that although the induced dipole phenomenon is the standard explanation of the London dispersion force, his 1930 model is significantly more sophisticated than this, involving detailed quantum mechanical calculations that go well beyond the concept of interacting dipoles.)

If a molecule already has a dipole—for example, water (H_2O) or hydrogen fluoride (HF)—then there is a *static* charge imbalance already present, and this will produce an additional dipole–dipole interaction over and above that produced due to the dispersion force. Similarly, in the case of ions—or ionic bonds, where electronic charge is transferred—rather than neutral atoms or molecules, there will be a strong electrostatic interaction by virtue of the charge on each species. If, in turn, it is covalent bonding, which involves the sharing of electrons, that dominates the interaction, we have to consider just how the electron probability clouds overlap and intermix in order to fully understand the quantum chemistry and physics.

Generally, a chemical bond is a mixture of all of these effects; very few are purely covalent or purely ionic, for example. Despite the complications, however, in all cases the electromagnetic force underpins the interaction and the overall shape of the energy-vs-separation curve is the same: the curve traces out what is known as a *potential well*, which defines the amount of energy we need to separate the atoms (or, equivalently, how strongly they interact) and whose minimum is located at the equilibrium bond length, where the net force is zero. Note both the similarity of the curves in Figure 3, and the variation in the position of the minimum of the force-separation graph (which represents the equilibrium separation of the atoms), in each case. (The experimental data were acquired using an atomic force microscope (AFM) and are taken from the work of Shigeki Kawai (International Center for Materials Nanoarchitectonics, Tsukuba, Japan) and collaborators.)

We've covered the '... attracting each other when they are a little distance apart...' clause of Feynman's pithiest scientific sentence. What about the '... repelling upon being squeezed into one another' aspect of the interatomic interaction? How does that arise?

Pauli pushes back

The attractive interaction that brings atoms together is ultimately due to the interaction of *unlike* charges. One might therefore imagine that the repulsion between two atoms (or molecules) when they move closer than their equilibrium separation is simply due to *like* charges repelling each other. That's indeed part of the repulsion, but it's far from the whole story. The fundamental reason that a pair of atoms or molecules strongly resists being pushed together to a separation smaller than their equilibrium bond length is, in fact, down to a fundamental rule of quantum mechanics: the Pauli exclusion principle.

Pauli's principle is a cornerstone of our entire universe. Without Pauli exclusion, we wouldn't have the Periodic Table of the elements, and matter would behave very differently indeed. It is ultimately the Pauli exclusion principle that's stopping you falling through your seat—or through the ground if you happen to be standing—as you read this sentence. The principle has its roots in quantum field theory—it requires a sophisticated combination of relativistic physics and quantum mechanics to be understood in even a reasonably complete fashion. Fortunately, we don't need to dig that deep for our purposes here. The Pauli exclusion principle can be stated rather more straightforwardly as follows: no two electrons can occupy the same quantum state. (The exclusion principle is, in fact, much broader in scope than this, because it applies not just to electrons but to an entire class of quantum particles known as fermions. As emphasized many times above, however, when it comes to nanotech our primary focus is on electrons.)

This means that, if you'll excuse the anthropomorphism, electrons will do their utmost to avoid each other if they have the same quantum properties (including, in particular, an attribute known as spin, which is responsible for (nano)magnetism and which we'll come to in Chapter 4). They are exceptionally antisocial entities when considered in this light. It is the exclusion principle that gives rise to most of the repulsion experienced by two atoms or molecules if they are pushed closer than their natural, stable separation. And it is the exclusion principle that provides the reaction force that stops you walking (or falling) through another object. Forces at the nanoscale have an influence that extends all the way to the macroscopic world around us (and beyond).

More than a theory

Although the theory of interatomic and intermolecular forces outlined above was supported by experimental evidence involving vast assemblies of interacting atoms or molecules (in the solid, liquid, or gas phase), it was only with the advent of the scanning probe microscope in the early 1980s that those interactions could be probed on an atom-by-atom basis. The SPM now routinely allows us to measure the forces between atoms and molecules with not just atomic resolution but sub-atomic precision. By 'sub-atomic' I mean not that we're probing the nuclear structure of the atom—remember that we're many orders of magnitude away from the energy scales typical of the forces experienced by particles inside the nucleus—but rather that we can resolve variations in force and energy down to a level much smaller than the diameter of an atom.

The remarkable level of precision that is now possible with state-of-the-art nanoscience is shown in the inset to Figure 3. The Lennard-Jones potential—and, of course, all other models of the interactions of atoms and molecules—can now be probed and exploited on an atom-by-atom basis, with control of the interatomic separation down to the *pico*metre level (10^{-12} m), a precision that is comparable to approximately a hundredth of an atomic diameter. The ability to map and manipulate matter with this exceptional level of detail is now routinely exploited in nanoscience and nanotech laboratories across the world. This has, in turn, enabled remarkable insights into the quantum world that underpins so much of nanotechnology—research that even just a few decades ago would have been considered as a *Gedankenexperiment* (thought experiment) with little chance of ever being realized in the lab.

Chapter 2
The quantum, confined

Until the advent of scanning probe microscopes we did not have direct control of the atomic and molecular architecture and interactions that form the bedrock of the nanoscopic world. Although sophisticated and elegant chemical reactions can be exploited to synthesize a wide variety of tailored nanostructures/ nanobjects—NanoKid from Chapter 1 is certainly a remarkable *tour de force* demonstration of the power of synthetic chemistry (and we'll be seeing other examples of the impressive capabilities of chemical synthesis very soon)—only scanning probe microscopes can manipulate matter in the manner shown in Figure 4.

That figure shows a few frames from what has been dubbed the world's smallest movie by its creators, a team led by Andreas Heinrich at IBM Research Labs (Almaden). (Fittingly, Heinrich—currently the Director of the Centre for Quantum Nanoscience (QNS) in Seoul, South Korea—was the successor to Don Eigler at Almaden. It was Eigler who, with his colleague Eric Schweizer, first manipulated atoms to create the IBM logo discussed in the previous chapter.) Each bright blob in the frames above is a single carbon monoxide (CO) molecule that has been manipulated into place using the tip of a scanning probe microscope. The frames were not just imaged but 'drawn', molecule by molecule, using the SPM. Operating in ultrahigh vacuum—at a pressure not too different from that in deep

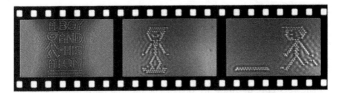

4. Frames from 'A Boy and His Atom', the world's smallest stop motion video, created by a team of nanoscientists at IBM Research Labs (Almaden) led by Andreas Heinrich.

space—so as to ensure that all possible contaminants were purged from the experiment—and at a temperature of roughly four degrees above absolute zero (to ensure that thermal energy did not jiggle the CO molecules out of place), Heinrich's team brought NanoKid's SPM-assembled counterpart to life using the traditional 'stop motion' movie-making technique.

Although 'A Boy and His Atom' is an impressive demonstration of the sophisticated control of the building blocks of matter possible with scanning probes, nanoscience, of course, is about more than fun, games, and viral videos. Our ability to fabricate nanostructures an atom at a time, exploiting and directing interatomic (and intermolecular) forces, means that we can build bespoke containers for electrons. And, for all of the reasons described in Chapter 1, once we can corral and confine electrons the way we want, we can define—even 'dial in'—the properties of materials in a way that was unimaginable before the advent of nanotechnology.

The control of electrons in this way is nothing less than fundamental quantum physics realized in the laboratory. The interface between quantum technology and nanotechnology is almost always ill defined; once we work at the nanoscale, we generally must take into account the quantum nature of matter and so the terms 'nano' and 'quantum' can often be used interchangeably. (Indeed, Heinrich's team is at the forefront of research at the quantum–nano interface.) Why is there this substantial overlap of quantum mechanics and nanotechnology?

Making waves: the nano–quantum interface

At the very core of quantum physics is the idea that matter, under the right circumstances, exhibits wavelike characteristics. This is not to say that matter—be it sub-atomic particles, atoms, molecules, billiard balls, or books—*is* a wave; it's not that electrons themselves, for example, spread out like ripples of water on a pond. Rather, it's that matter behaves as if it were a wave. And the extent to which the wavelike characteristics of matter are manifest depends critically on the size of the object. This is why nanotechnology and quantum physics are so often synonymous: at the nanoscale, quantum phenomena are ubiquitous.

In a remarkable leap of scientific creativity and imagination, the French physicist Louis de Broglie postulated (in his 1924 PhD thesis) that just as light, which clearly acts like a wave, can be considered as a stream of particles (i.e. photons), so too matter, whose atoms are traditionally thought of as 'billiard ball' particles, can behave as a wave. This was a revolutionary insight made all the more astounding for the simplicity of the equation that bears de Broglie's name, a cornerstone of quantum physics:

$$\lambda = h \,/\, p$$

In that equation, λ is the wavelength associated with the particular particle of matter in question, h is Planck's constant—another mainstay of quantum mechanics (about which we'll have more to say very soon)—and p is the particle's momentum. Matter doesn't behave like a wave in the macroscopic world around us—you don't diffract when you walk through a doorway, for example—because we're simply too large. Putting some numbers into de Broglie's equation above brings home the essential significance of size in the quantum world (although I must stress that the following is an oversimplified back-of-the-envelope calculation for illustrative purposes only).

Let's take a human of average mass, say 76 kg (this varies somewhat from country to country—I've chosen the UK value), moving at a walking pace of 1.5 metres per second. The amount of momentum associated with this average Jane/Joe is simply the product of their mass and their speed, that is they have a momentum, p, of 114 kgm/s. What's now essential to realize is that Planck's constant is a very small number indeed, $h = 6.63 \times 10^{-34}$ Js (where the units, Js, are joules seconds). If we plug these values of h and p into de Broglie's equation, we find that the wavelength associated with a typical human is truly tiny: 10^{-36} metres. That's roughly 29 orders of magnitude smaller than the wavelength of visible light.

A more striking comparison is to consider the size disparity between ourselves and the observable universe. The latter is approximately 10^{27} metres in diameter, whereas I stand 1.78 metres tall in my stockinged feet—a thoroughly average height. In other words, the difference in size of a human as compared to the entire observable universe is dwarfed by the disparity between our size and our de Broglie wavelength, a factor of 10^{36}. This is the fundamental reason why we don't see humans behave as quantum particles: we're just too big.

At the nanoscale, however, it's a very small world indeed. Let's repeat our simple calculation for an electron, which has a mass of 9.11×10^{-31} kg. What do we choose for the speed? Depending on the context, there are a wide variety of answers to 'What's the speed of an electron?' (It's a little like a quantum mechanical version of 'How long is a piece of string?') We'll choose a context appropriate for our focus on nanotechnology. Back in 2015, an international team of researchers headed by Reinhard Kienberger, Professor for Laser and X-Ray Physics at the Technische Universität München, determined that the speed of an electron travelling through an ultrathin layer of magnesium, just a few atoms thick, was approximately 5,000 km per second. If we take the product of the electron's mass and its velocity we find that its momentum, as compared to that of a human sauntering along at walking pace, is unimaginably small: 4.6×10^{-24} kgm/s. This means that the de Broglie wavelength of the electron in question is approximately 0.14 nm—comparable to the diameter of an atom. That matching of the electron wavelength to the atomic length scale is of key importance. Just as light waves diffract from objects comparable to their wavelength (giving rise to the rainbow of colours on the underside of a CD or DVD, for example), a beam of electrons incident on the surface of a crystal will be diffracted by the atomic lattice.

It was diffraction of just this type that provided the first, and exceptionally compelling, empirical evidence of the wave-like nature of electrons—and, by extension, all of matter. In a pioneering experiment carried out by two scientists, Clinton Davisson and Lester Germer, at Western Electric (which later became Bell Labs) a century ago, electrons were diffracted from a nickel crystal, forming a diffraction pattern on a fluorescent screen. Long before the invention of the scanning probe microscope, electron diffraction was used to determine just how atoms are arranged in materials. (This is not to say that SPM has superseded electron diffraction. Far from it—scanning probes and electron diffraction complement each other in the information they provide on the atomic, molecular, and nanoscale structure of matter.)

One could perhaps argue that a diffraction pattern is not direct evidence of matter waves; we see maxima and minima and interpret those as arising from wavelike behaviour, but perhaps the pattern has another source? There are many reasons for interpreting the diffraction pattern as arising from wave interference but the most striking evidence for the wavelike characteristics of matter again comes from scanning probe microscopy. Take a close look at those frames from the 'A Boy and His Atom' movie. Note how, in each case, the arrangement of CO molecules is surrounded by a pattern of ripples; the probe microscope is imaging the interference of electron waves. We can see the wavelike behaviour of matter right before our eyes on a computer screen.

What's even better is that the scanning probe microscope allows us to construct containers for the electron waves. Atoms and molecules can be arranged almost at will (subject only to the constraints of the interatomic and intermolecular forces at the surface of the sample) to reflect and confine the waves to a region of space. The iconic example of this type of confinement is the quantum corral shown in Figure 5, a nanoscale ring of atoms that was built to contain the electrons within. Just like the IBM logo in Chapter 1, the corral was constructed at IBM Almaden by Don Eigler's group (although this time the work was led by Mike Crommie, now at Berkeley, and it involves iron atoms on copper rather than xenon on nickel). I've also included an image of the standing wave formed in a gently shaken coffee cup for comparison. You may well have seen that characteristic pattern of concentric circles form in your coffee/tea cup while waiting for a train to depart a station. The vibrations of the carriage are transferred to the liquid, causing it to resonate, and the sloshing back and forth of the liquid forms a standing wave because the coffee is *confined* to the cup.

From many perspectives, it's very similar physics that gives rise to the pattern inside the quantum corral. The electron waves get reflected from the surrounding wall of atoms, bouncing back and forth, and interfering to form the standing wave pattern we see inside the corral. Despite the massive difference in length scale (nanometres vs centimetres), temperature (4 K vs room temperature, i.e. ~ 300 K), pressure (ultrahigh vacuum vs atmosphere), and material (solid copper vs liquid coffee), the pattern is exactly the same. (Indeed, it's mathematically exactly

5. Left: A quantum corral comprising 48 iron atoms, each positioned using the tip of a scanning tunnelling microscope (STM), on the surface of a copper crystal. The corral is approximately 10 nm in diameter and was fabricated by Mike Crommie and colleagues at IBM Research Labs in 1993. Striking evidence for the wave characteristics of matter at the nanoscale is visible within the ring of Fe atoms. Right: Standing wave formed in a vibrating cup of coffee. The pattern formed at the surface of the coffee is mathematically identical to that formed in the quantum corral.

the same type of pattern—something known as a Bessel function.) Remarkably, the only thing that the coffee-in-cup and electrons-in-corral systems have in common is the symmetry of the confinement: it's a circle in each case.

Despite the mathematical similarity, the standing wave inside the quantum corral is nonetheless of a very different form from its macroscopic caffeinated counterpart. In the coffee cup, the motion of the liquid produces real, physical waves. At the quantum level, however, the pattern inside the corral is a *probability* wave. The peaks and troughs represent, respectively, regions of high and low probability of finding an electron. The scanning tunnelling microscope (STM) is exquisitely sensitive to the presence or absence of electrons under the tip and so it maps out the variation in probability—technically, the probability density—right down to not just the atomic level but with a resolution that is much smaller than the size of an atom. Unlike the coffee, the pattern observed in the quantum corral is stationary: a wave frozen in time. Moreover, and despite all the talk of uncertainty, randomness, and unpredictability in popular science accounts of quantum physics, it represents a variation in probability that's perfectly predictable.

As you might imagine, if we change the shape of the container then the pattern of confined electrons adjusts to represent the new symmetry. What's arguably even more important, however, is that the *energies* of the electrons depend on the shape and size of the nanostructure that contains them. We can understand this best by once again considering parallels between the nanoscale and the macroscopic world. Quantum mechanics is, at core, simply a theory of waves. Therefore, much of our understanding of wave phenomena in the world around us can be scaled down to help interpret just how matter behaves at the nanoscale. We've already seen this for waves in a coffee cup. Let's consider an even simpler example: a guitar string.

When we pluck a guitar string we excite waves whose peaks and troughs remain at the same positions along its length (an example is shown in Figure 6): these are standing waves, as distinct from travelling waves, and are the one-dimensional analogues of the 2D standing waves seen in the quantum corral and the coffee cup of Figure 5. The characteristic sound of a guitar string comes from the mixture of the various standing waves (or harmonics) in which it can resonate. That the same note, say middle C, sounds different when played on guitar, piano, violin—or whichever stringed instrument you prefer—is because each of the instruments has its own signature mix of harmonics. The timbre and tone of all music—indeed, of all sound—ultimately depends on blends of harmonics.

Nanoscale waves and wavefunctions

What has this all got to do with nanotechnology and quantum physics? Everything. Just as when we shorten the length of a guitar string we increase the pitch of the resulting note, so too do we change the energy of electrons by confining their associated waves to a smaller region of space. The shorter the length of vibrating string, the higher the pitch; the smaller the nanostructure, the higher the electron energy. With an STM we can build nanoscale 'strings' to confine electron waves and so control their energy. Moreover, the electron energy is discrete, i.e. quantized, just like in an atom, except that the nanostructure may well comprise tens, hundreds, or even thousands of atoms. Although the Periodic Table gives us a wide selection of different atoms from which to choose, the electron energies are fixed in each case by the potential defined by the atomic nucleus. Nanotechnology instead allows us to construct *artificial atoms*: nanostructures that have discrete, tunable electron energy levels. Also variously called quantum dots, nanoclusters, nanoparticles, or nanocrystals, artificial atoms can be used to form tailor-made structures, devices, and materials, engineered from the bottom up.

An especially impressive example is shown in Figure 6, taken from work by Stefan Folsch's group at the Paul-Drude-Institut für Festkörperelektronik in Berlin. Folsch's team have built a one-dimensional 'string' from indium atoms and, using the same

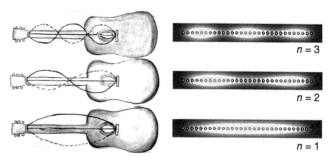

$n = 3$

$n = 2$

$n = 1$

6. There are very close parallels between the standing waves that form on guitar strings and the probability waves associated with electrons in nanostructures. The panels on the right in each case show a chain (or string) of 30 indium atoms, built using a scanning tunnelling microscope.

STM that created the nanostructure, have mapped out the probability waves for the confined electrons. Superimposed on the atomic string shown in Figure 6 are the variations in conductance—that is the flow of electrons between the STM tip and the chain—along its length. The conductance is directly proportional to the probability density—the brighter a region, the more likely it is for an electron to be found there. The pattern of peaks and troughs (or antinodes and nodes, in the language of waves) in the nanoscale string of Figure 6 is identical to that found for the various modes of vibration (otherwise known as resonances, harmonics, or eigenmodes) of a macroscopic string fixed at both ends. In other words, they see precisely the same standing wave patterns at the quantum level as are observed for a vibrating guitar string—direct visual, and visceral, evidence of not just the wavelike nature of matter but the remarkable effectiveness of mathematical physics in explaining such disparate systems at such dramatically different length scales.

Although there are very many similarities between the physics of the guitar string and that of its nanoscale counterpart, there's an essential difference that ultimately gives rise to much of the 'weirdness' of the quantum world. The electron waves on the nanostring are described by what is known in quantum mechanics as a *wavefunction*. Despite the wavefunction's central importance in quantum mechanics, it is not something we can observe directly in any experiment. This is ultimately because it involves complex numbers, which contain the imaginary quantity $i = \sqrt{-1}$. We write a complex number, z, as $z = x + iy$, where both x and y are real numbers. A square root of a negative number is not something we can measure in an experiment—it's not a real number. What we can measure, however, is zz^*, where $z^* = x - iy$ and is called the *complex conjugate* of z. If we take the product of a complex number, or a complex-valued mathematical function, with its conjugate, note that we eliminate the imaginary number, that is, $zz^* = x^2 + y^2$. In other words, zz^* is real and *is therefore measurable*.

A wavefunction, however, isn't just a complex number—it's a function of complex numbers, or, in short, a complex function. The wavefunction itself, which we traditionally represent by the Greek symbol psi, ψ, is not something that can be observed directly in experiment (because it is complex). However, the product of ψ and its complex conjugate, that is $\psi\psi^*$, is a real function—real in the mathematical sense—and *is* something we can observe in experiment. It is a function that describes how the probability of finding a quantum particle—in our case, an electron—varies in space and time. Remarkably, STM can directly image these probability maps. The technique produces images of $\psi\psi^*$—otherwise known as the probability density—and this is what we're seeing in Figures 5 and 6. (There are some provisos and nuances here with regard to the role of the STM tip itself, for one, but to a very good approximation, Figures 5 and 6 are maps of probability density.)

In Folsch et al.'s experiment (and similar earlier demonstrations by a number of other STM groups, including, in particular, Wilson Ho and colleagues at the University of California, Irvine), changing the size of the quantum string—even by the addition or removal of just a single atom—confines the electron wavefunction to a larger or smaller region of space. In this case it's for a one-dimensional nanostructure (the 'string'), but it could just as easily be a two-dimensional (as for the quantum corral) structure, or, as we'll see soon, a three-dimensional nanoscale object. The more tightly confined the electron—that is, the smaller the region of space in which it is free to move—the higher its kinetic energy.

This can also be interpreted in terms of the famous Heisenberg uncertainty principle. A more tightly confined electron means that its probability wave is less spread out in space. In other words, the range of possible values of its position is relatively small, meaning that, via Heisenberg's principle, the possible range of values of its momentum is correspondingly higher. Those higher momentum values are in turn associated with higher electron energy. (Note that this type of uncertainty is a fundamental property of waves of any description, at any scale, and not simply a result of measurement. Too often, the Heisenberg uncertainty principle is incorrectly reduced to a simplistic 'measurement disturbs the system' description. That issue, known as the measurement problem, is certainly a core aspect of quantum mechanics but it is very much distinct from the Heisenberg uncertainty principle.)

The level of control we now have at not just the nanoscale or atomic, but single chemical bond, level is astounding. With a scanning probe microscope we can build nanostructures an atom at a time, 'dialling in' the electron energies we require simply by controlling the size of the structure. And as highlighted in Chapter 1, once we can control the positions and energies of electrons, we can define and tune virtually whatever material property we like. Or, as elegantly pioneered by Michelle Simmons's group at the University of New South Wales (more of whom in Chapter 4), we can use SPMs to build working nanoelectronic devices that are atomically precise: to fabricate single atom transistors, to explore whether Ohm's law works all the way down to the atomic level (it does), and to define precisely where individual impurity atoms are placed in a semiconductor to control its conduction. But there's a big problem with the scanning probe microscope's unparalleled ability to map and manipulate the ultrasmall: it's exceptionally slow.

Serial to parallel

The quantum confinement nanostructures we've covered thus far have each been fabricated in what might best be described as an extreme environment: temperatures just a few degrees above absolute zero, in ultrahigh vacuum, and with samples whose surfaces have been prepared so that they are almost entirely free of contaminants, right down to the atomic level. Those types of constraint make translation of prototypical nanodevices into the less forgiving environs of the everyday world exceptionally challenging—no one wants to have to cool their phone, laptop, or tablet to four degrees above absolute zero before it works.

Equally problematic is the question of fabrication time. Another simple back-of-the-envelope calculation highlights just how the limited speed of today's SPM technology, especially when it comes to probing and positioning atoms one at a time, is a show-stopper when it comes to mass production of devices. Let's take a rough, optimistic estimate of the time required to carry out each atomic manipulation operation: we'll choose one second per atom. There are roughly 10^{14} atoms per centimetre at the surface of a silicon crystal. (Other materials will have slightly different surface atom densities but will fall within the range 10^{13}–10^{15} cm^{-2}.) Therefore, it will take a scanning probe microscope of the order of 10^{14} seconds to assemble a single atomic layer having an area of one square centimetre, if we 'pick and place' every single atom. That's a long time. A very long time.

All of recorded human history represents a timespan of roughly 5,000 years, that is approximately 1.6×10^{11} seconds. In other words, it would take an SPM working flat out, 24 hours a day, 365¼ days per year—with perfect, error-free single atom positioning—about 1,000 times longer than all of recorded history to fabricate a single atomic layer. Even if we were to see a million-fold increase in the speed of the technology so it takes a microsecond to put an atom in place, this still represents a total time of 10^8 seconds. That's over three years for a single, postage-stamp-sized atomic layer—hardly the most attractive proposition for a viable manufacturing technology.

Not all nanoscience is carried out under such uncompromising and inefficient conditions, however. Nature can assemble countless crystals comprising hundreds of millions of atomic layers routinely on a timescale of minutes or less, without any need for the atoms or molecules to be carefully directed into place by an external force like the SPM. Instead, the interatomic (and/or intermolecular) forces described in Chapter 1 bring the constituent atoms/molecules together in just the right way to minimize the total energy—a process known as self-assembly. Entropy, which is related to the total number of possible atomic and/or molecular configurations, also plays a big role in the assembly process and we'll see more about this in the next chapter.

We've already seen that synthetic chemistry involving liquid phase reactions in a much more traditional sample environment (i.e. test tubes and beakers) can be harnessed to yield nanostructures like NanoKid, with a phenomenal level of control over their atomic structure. This is self-assembly in action. Instead of atoms and molecules being moved excruciatingly slowly by 'brute force' with a scanning probe microscope, interatomic and intermolecular forces are instead exploited—and tuned, via subtle (and sometimes not-so-subtle) chemical modifications—to assemble nanostructures and microstructures. As we'll also see in the next chapter, self-assembly can be used to produce a dizzying array of nanostructures, of varying size, shape, and symmetry, and for which function and form are intrinsically linked.

NanoKid is only one very recent example of the power of 'wet' synthetic chemistry and self-assembly in nanotechnology. Almost 200 years ago, Michael Faraday—who could arguably be described as the first nanoscientist—pioneered the study of gold nanoparticles. (Faraday, however, did not use that term to describe the particles he synthesized and studied, and rather dismissively attributed the remarkable effects he observed to 'a mere variation in the size of particles'.) But 300 years before Faraday's experiments, Renaissance potters in Italy (and elsewhere) were unknowingly exploiting nanoparticles to produce coloured glazes in pottery. And 2,000 years before that, in 800 BC, nanoscale structure was ultimately, and inadvertently, responsible for the purple hue of Egyptian gold-plated ivory. From this perspective, nanotechnology is as old as science itself.

All that glitters...

I have been occupying myself with gold this summer; I did not feel headstrong enough for stronger things. The work has been of the mountain and mouse fashion; and if I ever publish it and it comes to your sight, I dare say you will think so:—the transparency of gold—its division—its action on light.

The quote above is taken from a letter that Faraday sent to his friend, the German chemist Christian Friedrich Schönbein, in early 1856 and is remarkable for its degree of self-effacement. Given that the work he is describing not only was a pioneering study of just how light interacts with nanostructured matter but also involved the controlled synthesis of nanoparticles for the first time, Faraday was clearly being rather too hard on himself. Although much better known for his pioneering work on electricity and magnetism, Michael Faraday in essence founded the entire field of colloidal chemistry—and, by extension, laid the foundations of nanochemistry—while studying the optical properties of suspensions of sub-microscopic gold particles in water.

Faraday's methods, or adaptations thereof, are now a standard approach to the synthesis of gold nanoparticles and are sufficiently straightforward that even a chemically inept physicist like myself can carry out the synthetic steps. In order to produce nanoscopic particles Faraday created a colloidal suspension—a dispersion of one phase in another, in this case (insoluble) solid gold clusters in water. This type of colloidal gold suspension can be formed by mixing standard off-the-shelf chemical compounds—gold chloride, sodium hydroxide, and citric acid—and can therefore be carried out by high school students. One remarkable aspect of this synthesis is the stability and longevity of the resulting colloidal suspension. It can take many years for the gold particles to aggregate and 'drop out' of solution; the charge of each stabilizes the colloidal suspension and keeps the nanoparticles from getting too close to each other. In a fridge in one of our labs there are a couple of containers of gold nanoparticle suspensions made by A-level students during a summer school more than a decade ago—they look just like they did on the day they were synthesized. More impressively, one of Faraday's own original nanoparticle suspensions—now 180 years old—is kept at the Royal Institution in London and has retained its ruby red colour for nearly two centuries.

Ruby red? Surely the colour of gold is gold? Not at the nanoscale. When gold is reduced in size from a bulk crystal to nanocrystals just a few nanometres (or a few tens of nanometres) across, it loses its characteristic hue and lustre and instead becomes deep red in colour. It was this dramatic change in colour, and the implications for the interaction of light with 'fine grained' matter, that fascinated Faraday. It was only with the advent of quantum mechanics in the early 20th century, however, that the physics underlying this colour conundrum could be fully understood. Once again, it is the wave characteristics of matter that are responsible for the dramatic change in the colour of gold.

I've thus far glossed over an important aspect of the waves seen inside the quantum corral of Figure 5: those patterns won't form on just any surface. Crommie and colleagues chose that particular copper surface with care because the electrons there are essentially free to roam. In the absence of any constraint—such as the ring of iron atoms that forms the corral, or contaminant atoms/molecules, or crystal defects—they can move across the entire crystal largely unimpeded. Because of this electronic freedom, physicists refer to the system as a *free electron gas* (or, in honour of the physicist who contributed so much to our understanding of the behaviour of matter, a *Fermi electron gas*). This gas—also sometimes referred to as a 'sea'—of electrons is a feature of many metal crystals, including copper and gold.

The freedom of the electrons to roam the gold crystal means that they respond collectively to a disturbance or an excitation such as an electromagnetic wave, that is visible light. Shine light on a macroscopic lump of gold (or copper, or silver, etc.) and its electrons slosh back and forth in synchronized motion, driven by the rapidly oscillating electric field of the light. Moreover, and just like the quantized energies of an electron confined to a nanostring, this 'sloshing' of the electrons is associated with discrete energy steps. The term given to this quantized oscillation is a plasmon. If the frequency, f, of the incident light (which is related to its wavelength, λ, via $c = f\lambda$, where c is the speed of light) is close to the rate of oscillation of the electrons, then we have what is known as a plasmon resonance.

Resonance is a phenomenon that crops up repeatedly across all science and engineering, regardless of system or scale. Wine glasses that resonate at the same pitch as a particular musical note (until they're driven to explode), bridges that collapse because their resonant frequency matches that of the driving force (be it due to the wind or synchronized footsteps), and children on swings who go higher and higher when energy is injected at just the right rate (but without, hopefully, either collapsing or exploding)—each is a resonant system which has its maximum response at a particular frequency. The electrons of the gold nanoparticle similarly resonate when the right driving frequency is reached. By, once again, simply changing the size of the nanostructure—in this case, a three-dimensional nanoparticle rather than a 2D quantum corral, or a 1D 'string'—the plasmon resonance can be tuned. The electrons are increasingly confined as the nanoparticle is made smaller, limiting the volume in which they can oscillate and shifting the resonance to higher frequency.

The colour of the gold nanoparticle is directly related to the frequency of the plasmon resonance. This determines the wavelength of light that will be absorbed most strongly. For 30 nm gold nanoparticles, for example, the plasmon resonance causes enhanced absorption of light in the blue-green region of the visible spectrum, around a peak wavelength of ~ 450 nm. Red light, which is of a much lower wavelength (~ 650–700 nm) is instead mostly reflected, giving rise to the characteristic ruby colour that so fascinated Faraday and which is the signature, once again, of the wavelike characteristics of matter at the quantum and nanoscale levels.

It's not just plasmon resonances of nanoparticles that exhibit this size dependence—a good rule of thumb is that the smaller an object, the higher the frequency at which it will resonate. 'Twang' a ruler at the edge of a table or desk and we know that the pitch of the resulting note depends on the free length that is oscillating. As the amount of the ruler overhanging the table is made shorter, the vibrational frequency gets higher. Keep reducing the size of an object to the nanoscale and its mechanical resonant frequency is not the tens of hertz of a vibrating ruler, or the hundreds of hertz of a resonating wine glass or tuning fork, it's instead of the order of millions of hertz (MHz) or more. That's a frequency scale more usually associated with a different region of the electromagnetic spectrum than visible light: radio waves. Remarkably, nanoscale objects can vibrate, mechanically, at comparably high rates to that of the oscillating electric field associated with a radio wave.

In 2019, scientists at Lancaster University and the University of Oxford directly measured the mechanical oscillation of a carbon nanotube, just 3 nm in diameter, suspended between two metal contacts. In other words, they fabricated a direct nanoscale analogue of a free-standing guitar string. They found that the carbon nanotube (about which we'll have more to say in Chapter 5) resonated at a frequency of 231 MHz—a note so impossibly high that it is inaudible to every organism on Earth, including those with the highest frequency thresholds: dolphins, bats, and the wax moth, whose hearing extends to 160 kHz, 250 kHz, and 300 kHz, respectively.

At the nanoscale, oscillation and vibration are everywhere: thermal energy at room temperature is enough to shake atoms, molecules, and nanostructures so that they are in constant motion. But even at the lowest temperature ever achieved, roughly 500 nanokelvin—more than a million times colder than the temperature of deep space, 2.7 K—there is still vibration at the nanoscale. We can never freeze out this vibration because to do so would violate the Heisenberg uncertainty principle: zero vibrational motion would mean that the position of an object was completely defined, meaning that its momentum spanned an infinitely wide range of possible values.

Artificial atoms: zero dimensional nanoscience

Metals have underpinned humanity's technological development since well before antiquity—prehistory is defined in many ways by the exploitation of various metals, not least, of course, in the Bronze Age and Iron Age. It's perhaps not surprising, therefore, that metals continue to play a central role in 21st-century technology—some have even referred to the intense interest in gold nanoparticles as 'the second Gold Rush'. The plasmon resonances of metal nanoparticles discussed above are exploited not just in nanoscale optics and optoelectronics, but are used throughout bionanotechnology and nanochemistry in a variety of contexts including sensing mechanisms, catalysis of chemical reactions, and as nanoscopic antennae. In particular, when molecules—including large biomolecules like proteins and antibodies—bind to the surface of a metal nanoparticle, they disturb the collective oscillation of the electrons and thus change the frequency and intensity of the plasmon signal. In the best cases, this modification of the plasmon, in concert with other spectroscopic techniques, can be used as a type of molecular fingerprint, identifying the particular chemical species that have bound to the surface of the particles.

This barely begins to scratch the surface of the applications of metal nanoparticles in nanoscience, and we'll return to a few more examples later. When it comes to information technology, however, it's not metals but semiconductors that rule the roost. Silicon—rather than, say, silver, sodium, or steel—has driven the microelectronics industry from its outset back in the early 1960s right up to its 21st-century nanoelectronics counterpart. Why is this?

Semiconductors have a key advantage over metals when it comes to electronics: their conductivity is switchable and tunable. Traditionally, metals are 'always on' when it comes to electron flow, whereas the dominance of semiconductors in solid state electronics and, by extension, the ICT (information and communication technology) industries, is because their conductivity can be easily modified and switched off (and back on again). The ability of a material to conduct electricity—or not—arises from just how the energies of its constituent electrons are distributed. We can understand why this is the case by first starting with two atoms and then progressively adding more, building a nanoparticle from the bottom up and considering how the arrangement of electrons evolves with size.

Remember the interatomic potential described in Chapter 1? In that case we assumed that the atoms interacted rather weakly via fluctuations of their constituent electrons. In order to understand the electronic properties of materials we need to go beyond those rather weak, if ubiquitous, interactions and consider just what happens when chemical bonds are formed. When two atoms interact via van der Waals forces, there's really no bond formed—it's what is known as a physical, rather than chemical, interaction. (For an atom held in place on a surface in this way—as for the IBM logo formed by xenon atoms on nickel—we say the atoms are physisorbed rather than chemisorbed.) When a covalent bond forms, however, the electrons involved are not confined to their parent atom: their quantum mechanical probability cloud spreads out over the resulting molecule. This has key implications for how the molecule—and, as we add more atoms, the nanoparticle, and finally the solid (which we can think of as a very large molecule)—conducts electricity.

A schematic diagram (Figure 7) speaks a thousand words here. For simplicity and clarity we'll consider hydrogen, with its one lonely electron. For each isolated H atom in Figure 7(a), the probability of finding the electron in a particular region of space around the nucleus can be determined from a mathematical function known as an atomic orbital. An atomic orbital is, in essence, a one-electron wavefunction. As for the electrons confined to the corral (Figure 5) or the 'nanostring' (Figure 6), the atom represents another form of quantum confinement. In this case, the nucleus establishes the electrostatic potential that confines the electron.

The bonding orbital has a lower energy than that associated with the atomic orbital of each isolated hydrogen atom. This is the driving force for the formation of the chemical bond in the first place: the system reaches a lower energy via the overlap of the atomic orbitals. Both electrons (one from each of the hydrogen atoms) occupy the bonding orbital. As also sketched in Figure 7(a), the bonding orbital results in a high probability of the electron being found between the protons: the electronic charge acts as a 'glue' holding the atoms together. The antibonding orbital, however, is associated with a very high probability for the electron to avoid the region between the protons. As its name suggests, if an electron finds itself in the antibonding orbital, the chemical bond is destabilized and the molecule can dissociate back into its component atoms.

The antibonding orbital is also higher in energy. This means that the 'natural', lowest energy state of an electron—in quantum mechanics we call this the ground state—is when it is found in the bonding orbital. In order for an electron to occupy the antibonding orbital, an injection of energy is needed. This can come in the form of thermal energy (i.e. heating) or photons: a

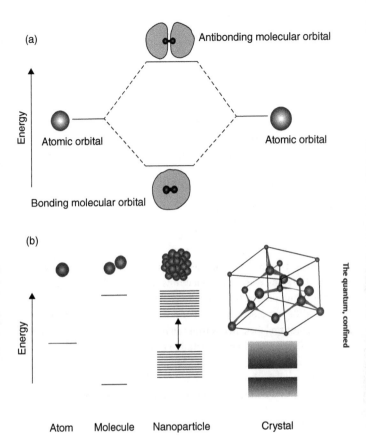

7. From energy levels to bands. (a) Two hydrogen atoms, each with a single electron in a 1 s atomic orbital, are brought together to form a hydrogen molecule. This leads to the formation of molecular orbitals— one, the bonding orbital, resulting from the constructive interference of the atomic orbitals, and the other, the antibonding orbital, arising from destructive interference. (b) Evolution of energy levels to form energy bands. As the number of atoms increases from left to right so too do the number of energy levels, to the point where, for a sufficiently large chunk of matter, the levels are no longer discrete and form effectively continuous bands.

molecule can be dissociated by choosing light of the right energy to excite electrons from bonding to antibonding orbitals. It is exactly this type of photodissociation that is responsible for the formation of the ozone layer: oxygen molecules are split up into their component atoms. The absorption of light (or other forms of energy) doesn't necessarily have to break up a molecule completely, however. Instead, the input of energy can reorganize the bonding, once again by changing how the electrons populate different orbitals. That type of photochemistry has been exploited in many areas of nanotechnology to fabricate nanostructures.

Magic nano numbers

As we saw in Chapter 1, the Pauli exclusion principle tells us that no two electrons can be in the same quantum state. That doesn't mean, as is sometimes misleadingly suggested, that they can't have the same energy. Two electrons can indeed have the same energy and occupy the same molecular orbital, as long as they have different electron *spin*. For now, we're just going to treat spin as a quantum label: an electron can be either spin-up or spin-down. We'll discuss spin in more detail in Chapter 5 in the context of spintronics—a key component of nanotechnology—but for the moment, all we need to know is that spin is a property of quantum particles that defines just how they interact with each other.

We're familiar with the idea of charge controlling whether two particles attract or repel. This is something we generally take for granted because we learn it at a relatively early age, often via a party trick involving a balloon rubbed against our hair. But ask yourself this: where, fundamentally, does charge come from? It is not at all an easy question to answer. Like charge, spin is a property of electrons—and other quantum mechanical particles or objects—whose fundamental origin is rather deep in the bowels of fundamental physics but whose effects can be reduced to very simple rules. Just as we can condense this complex fundamental physics to 'like charges repel, dislike charges attract', so too can we treat spin as nothing more than a label on a particle.

Returning to the hydrogen molecule example, we now understand just how, and why, atoms form covalent bonds: the bonding produces a molecular orbital with a lower energy than the atomic orbitals. The electrons prefer to be in their ground state so they both occupy the lower energy bonding orbital. One electron is spin up, the other spin down, as represented by the arrows in Figure 7(a)—this is possible because although the electrons have the same energy, they're not in the same quantum state (due to the difference in spin).

The question now arises: what happens if we add more atoms? What if we keep adding atoms so that we first build a nanoparticle, and then keep going so we make a solid—a macroscopic piece of matter visible to the naked eye? The Pauli exclusion principle means that all of the electrons can't just collapse into the same quantum state. Paul Ehrenfest (1880–1933), an Austrian-Dutch physicist who made major contributions to both statistical mechanics and quantum mechanics, was the first to realize that, ultimately, it is the Pauli exclusion principle that underpins the stability of all matter in the universe. Without the exclusion principle, atoms would be compressible 'all the way down'. In other words, it's Pauli's principle that ultimately sets the length scale of matter: atoms and molecules are nanoscale objects because of the fundamentally antisocial nature of their electrons due to the exclusion principle.

Although it's the simplest possible element because of its lone electron, hydrogen is substantially more complicated when it comes to attempting to attach more atoms to the H_2 molecule. A hydrogen atom is very reactive—as we've seen there is a strong driving force for its lone electron to find a partner. This means that hydrogen readily forms stable H_2 units. Gaseous hydrogen—unlike noble gases like xenon—doesn't exist as a collection of atoms, because of their high reactivity. Instead, hydrogen gas comprises H_2 molecules. These can in turn interact to form solid hydrogen at sufficiently low temperatures (14 degrees above absolute zero) but they remain as distinct units, held together by the weak London dispersion forces described in Chapter 1. If, however, we turn up the pressure then it has been predicted theoretically that hydrogen could well form a lattice: a solid phase where the hydrogen atoms form an ordered, extended crystalline lattice held together by the electron 'glue'. However, this involves *very* extreme conditions, including pressures that are millions of times higher than atmospheric pressure, in order to 'encourage' the hydrogen atoms out of their H_2 state.

Fortunately, a wide variety of other elements, including gold and silicon, form crystal lattices rather more readily. Let's consider what happens when we add additional atoms to form a trimer (i.e. a three atom nanocluster), then a tetramer, then a pentamer...all the way up to a nanoparticle comprising hundreds of atoms. The electrons can't all occupy the same quantum state so as we increase the size of the cluster, new energy levels appear so as to accommodate the growing number of electrons. This is shown in Figure 7(b). The more atoms we add, the more electron energy levels appear. This is very similar indeed to the evolution of the electronic structure for the elements in the Periodic Table except that we're not increasing the size of a nucleus by adding more protons and neutrons—we're adding entire atoms to the cluster.

For gold, nanoparticles smaller than 12 atoms tend to be two-dimensional structures (somewhat like the 2D 'string' of indium atoms shown in Figure 6). However, particles larger than this form 3D, 'quasi-spherical' structures that are more stable when the total number of valence electrons in the cluster—that is the most loosely bound electrons involved in bonding—equals a certain 'magic' number. In other words, some clusters are much more stable than others because they've got just the right number of electrons. This is precisely the same phenomenon that underpins the entire Periodic Table, except in that case it's individual atoms, rather than multi-atom clusters, whose enhanced stability arises from their having the right number of electrons.

The fundamental reason we have columns in the Periodic Table is because, metaphorically, the electrons in atoms 'stack' in shells with only so many being allowed in each shell. This shell model is, like Bohr's picture of the atom, a fiction, but it's nonetheless a very helpful notion to keep in mind, not least because it explains so much of chemistry. In particular, the nobility of the noble gases stems entirely from the completeness of their electron shells; their atomic numbers, that is 2 (He), 10 (Ne), 18 (Ar), 36 (Kr), 54 (Xe),... reflect this. Those numbers are magical in the sense that when an atom has precisely that number of electrons, it is much less chemically reactive than its neighbours on the Periodic Table. The magic number effect seen for gold (and other) nanoparticles is remarkably similar: when a shell of valence electrons is complete, the particle stability is dramatically enhanced. This is one reason why nanoparticles are often referred to as artificial atoms (or designer atoms): the cluster as a whole behaves just like a giant atom. In principle, we could design a new Periodic Table in which we control the electronic structure of nanoparticles, and thus their chemistry, by adding or removing atoms.

Magic numbers for nanoparticles don't only have to be electronic in origin, however. There is also enhanced stability for what are known as geometric magic numbers, where the structure of the particle is such that there are just enough atoms to ensure a low energy arrangement that minimizes surface energy. Atoms are rather gregarious by nature and tend to prefer being surrounded by neighbours rather than being isolated. This gives rise to enhanced stability for complete shells of not just electrons but atoms themselves.

Mind the gap

When we add atoms to a nanoparticle, it's not just the bonding orbitals that are affected—the number and spacing of the antibonding orbitals also change. It would take a lengthy detour into undergraduate quantum mechanics and solid state physics/chemistry to fully explain just why this happens but, fundamentally, the various electron waves interfere, both constructively and destructively, in different ways as the nanoparticle increases in size. This changes how the orbitals are distributed in energy. Just as for the hydrogen molecule, the nanoparticle's bonding orbitals are filled with electrons, whereas the antibonding orbitals are empty. This means that there's an energy gap between the filled and empty states. As we continue to add atoms to the nanoparticle, the number of energy states increases, as sketched in Figure 7, because the Pauli exclusion principle prohibits all electrons from being in the same quantum state.

By the time we've added enough atoms to produce a mole of a material, that is 6.022×10^{23} atoms, there are so many energy levels for the electrons that we can no longer really consider those levels as discrete states. They instead form bands, as also illustrated in Figure 7. In the case of a semiconductor like silicon, there is a valence band filled with electrons (which originates from the bonding orbitals) and a conduction band that is empty, but can accept electrons. The energy gap between these bands—the *band gap*—determines not just how the material conducts electricity but also fundamentally underpins its optical properties. Photons of light can excite electrons from the valence band to the conduction band, but only if they have sufficient energy to 'bridge the gap'.

There's a simple, inverse relationship between the wavelength of light, λ, and the photon energy, E:

$$E = \frac{hc}{\lambda}$$

where c is the speed of light. Shorter wavelengths are associated with higher energy. If the photon energy is smaller than the band gap energy, the electron can't be excited and, save for reflections at the surfaces of the crystal, the light will be transmitted through the semiconductor. If we steadily decrease the wavelength of the incident light, however, there comes a point when the photon energy first matches, and then exceeds, the band gap energy. This means that the light can now excite electronic transitions and is therefore absorbed by the semiconductor crystal, instead of making it through to the other side. Each photon, that is each packet of energy, is consumed when it excites an electron to the conduction band.

Silicon has a band gap energy of 1.1 eV, whereas visible light spans a range of ~ 1.5 eV (for photons of red light) to 3.0 eV (for the blue end of the spectrum). This means that silicon absorbs the entire visible spectrum—again, leaving aside the issue of reflections from the surface—and so is an opaque crystal, appearing dark grey in colour. Silicon dioxide, however, has a much larger band gap, typically of order 9 eV (and so is described as an insulator, rather than a semiconductor). This far exceeds the energy of photons in the visible spectrum and so silicon dioxide doesn't absorb light of that energy. This is why glass, whose major constituent is silicon dioxide, is transparent.

As the size of a nanoparticle is reduced, the band gap widens due to quantum confinement. For semiconductor nanoparticles, it is possible to tune the wavelengths of light they absorb simply by changing their size. This is a remarkable demonstration of the power and versatility of nanotechnology: changes in size alone are enough to dramatically, and very visually, alter the properties of a material. Nanotechnologists now routinely harness quantum confinement in optoelectronics applications ranging from nanoparticle-based solar cells to high definition TV. That preceding sentence does not, however, begin to do justice to the long process that accompanies the commercialization of any innovative hi-tech device. There is an exceptionally tortuous route from a research lab prototype of a nanodevice to its successful adoption in the wider world.

Chapter 3
Tearing it down, building it up

'Tear Up The Books, Kids. Little Daisy (ENIAC, for short) Is Going To End Math.'

So ran the headline in the *Philadelphia Record* on 19 February 1946. ENIAC, the Electronic Numerical Integrator And Computer, financed by the US Army and developed by John Mauchly and J. Presper Eckert of the University of Pennsylvania, was an astounding accomplishment. Designed to be the first programmable, general-purpose digital computer, ENIAC contained more than 17,500 vacuum tubes connected by half a million soldered connections, and was enormous. Weighing in at almost 50 tons, it dwarfed its predecessor, Colossus—which was developed at Bletchley Park, the top-secret site of Second World War code-breaking—by quite some margin. (The Alan Turing-inspired Colossus, used to decode encrypted German teleprinter messages, shortened the Second World War by many months, saving thousands of lives.) While Colossus required a space the size of a living room, ENIAC was considerably bigger than a house: it occupied 167 square metres, as compared to the 68 square metres of living space on average that is now offered by modern UK homes.

Newspaper headlines across the world echoed the *Philadelphia Record*'s excitement about ENIAC—it was described as a super-brain, a lightning-fast robot computer, and a mechanical mathematician capable of out-thinking Einstein. Its performance certainly was exceptionally impressive for the time: ENIAC could achieve a blisteringly fast 5,000 instructions per second. In comparison, a smartphone now routinely computes at a rate of *billions* of instructions per second. Even if you don't have the latest model in your pocket, the speed of the processor powering an iPhone has been at the 1 GHz level (i.e. roughly a billion instructions per second) since 2010.

Miniaturization and speed of processing go hand in hand; the smaller our technology gets, the faster it becomes. The processors at the core of our phones, laptops, and tablets pack an astonishing number of electronic components into a remarkably small space. In terms of component size alone, microelectronics gave way to nanoelectronics quite some time ago. At the dawn of the semiconductor industry in the late 1960s, integrated circuits involved miniaturization of components down to the tens of micron scale (i.e. tens of thousands of nanometres). Skipping forward a few decades, the Intel 80386 processor that powered the PC on which I wrote the code for my final year undergraduate project back in 1990 had a feature size of 1,000 nm. (In the semiconductor device industry, feature size historically used to refer to the length of the channel between what are known as the source and drain connections of a transistor. Nowadays, feature size is taken to mean the smallest structure on the transistor.) Intel's Pentium 4 chip, released in 2004, broke the 100 nm barrier via the so-called 90 nm process. Fifteen years later, in December 2019, Intel announced plans for 1.4 nm production by 2029.

That's a reduction in feature size of roughly five orders of magnitude—a staggering level of enduring technological ingenuity that was predicted by Gordon Moore, the former CEO (and co-founder) of Intel long before nanotechnology was imagined. Moore's law posits that the number of transistors in an integrated circuit doubles approximately every two years. Although always described as a law, there is nothing fundamental about Moore's prediction—it's not in the same league as, for example, Newton's laws or the first, second, and third laws of thermodynamics; it's not directly derived from physics or chemistry principles. Instead, the underlying reasoning is based on economic arguments. Nonetheless, year-on-year increases in computational capability since the late 1960s have been in line with what might be better described as Moore's heuristic. Nanotechnology has played an essential role in ensuring that continued progress, but, ironically, and for reasons we'll get to soon, nanoscience is also why Moore's law is slowly dying.

Top-down technology

Semiconductor fabrication plants, or foundries, are exceptionally sterile environments. Devices are fabricated in pristine clean rooms whose environmental conditions are carefully regulated to minimize contamination due to airborne particulates, and to control temperature, air pressure and airflow, humidity, vibration, noise, and lighting. This level of environmental management is not for the benefit of the semiconductor nanotechnologists working therein. Instead, clean rooms are essential in order to protect nanostructured devices from even the tiniest speck of dust. The sub-10 nm feature size in modern semiconductor technology means that even an invisible dust particle, which is typically thousands of nanometres in size, is gigantic by comparison and could be responsible for significant damage during the processing of the device.

How are the nanoscale components that underpin practically all digital technology created? Device fabrication involves a process that has been in place—but of course has steadily evolved—since the start of the semiconductor industry. It's fundamentally a 'top-down' lithographic process that involves transfer of a pattern into the surface of a silicon wafer. The word *lithography* comes from the combination of the Greek *lithos*, meaning stone, and *graphia*, meaning to write; literally, 'writing on stone'.

Twenty-first-century nanolithography involves writing on a silicon wafer—typically 300 mm in diameter for modern semiconductor fabrication plants—by a lengthy series of processing steps that remove unwanted material, defining the nanoscopic components by etching away the surrounding silicon. An overview of the key steps in this process, known as photolithography because of its reliance on light, is shown in Figure 8. It's important to realize, however, that this is a highly simplified schematic; a wafer could well go through variants of the photolithographic process up to 50 times during the fabrication of an integrated circuit.

Moreover, although photolithography is the industry standard, it is often complemented by other top-down techniques during the design and prototyping phases of chip development. Foremost among these is the use of a focused ion beam (FIB) to ablate

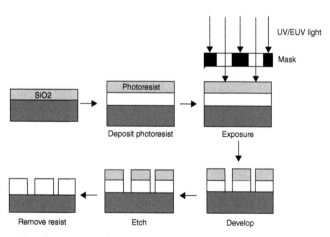

8. **Top-down semiconductor processing: nanolithography. The key steps in the patterning of a silicon wafer to form micro- and nanoscale features and components.**

matter right down to the nanometre level. Essentially, FIB is a nanoscale chisel or milling machine. A high energy beam of ions, usually gallium, is focused using electrostatic lenses onto a region of interest on a sample or a device, sputtering away material in a localized region. By rastering the beam back and forth, a wide variety of patterns can be written into the sample. In addition to its role in the semiconductor industry, the FIB technique is widely used for the preparation of samples for transmission electron microscopy (which, like scanning probe microscopy, is capable of atomic resolution, albeit via a rather less direct route). These need to be so thin—typically ~100 nm—as to be transparent to high energy electrons; FIB is a powerful tool for thinning samples to the required level.

A clean silicon wafer, with its protective layer of silicon oxide, is first covered with a layer of a light-sensitive material known as photoresist. Photoresists are a technology all of their own—an entire 'Very Short Introduction' could be written about the materials science, chemistry, and physics underpinning resists. In brief, illumination with ultraviolet light changes the chemical structure of the photoresist in the areas where it is exposed, reorganizing bonds and making the material soluble, or insoluble, in a chemical known as a developer. (There are many parallels between photolithography and traditional 'analogue' photography.) If the developer strips away the regions that have been illuminated, the resist is described as positive; if, instead, the developer removes the unexposed areas, it's a negative photoresist.

In order to select the regions where the resist will be exposed, a photomask is used. This is, in essence, a map of the circuitry and connections for the integrated circuit in question and defines whether the incident UV light is blocked or impinges on the resist-covered silicon wafer. UV light is used because the shorter wavelength—typically 193 nm in the majority of photolithographic processes at the time of writing—allows for better feature resolution. Remarkably, and despite the 193 nm wavelength, it is possible to fabricate sub-10 nm features using photolithography via sophisticated multiple pass techniques where the wafer is first patterned and then re-exposed using a different mask that is offset from the original, followed by another stage of resist development and removal.

After processing, the pattern encoded in the photomask has been transferred to the silicon wafer by virtue of the removal of oxide in specified regions. Silicon oxide, a wide band gap insulator, plays a central, essential role in the operation of microelectronic and nanoelectronic devices, equally important as the metal electrodes and silicon substrate that form the other ingredients required for what is known as complementary metal oxide semiconductor (CMOS) technology. CMOS has been the driving force of the industry for decades and for good reason: silicon oxide is chemically exceptionally stable and its wide band gap provides the high degree of isolation and insulation required to ensure that the vast numbers of individual components in an integrated circuit are decoupled from each other.

But despite all of this nanolithographic ingenuity driving miniaturization relentlessly forward (and downward) for decades, Moore's law is dying. In fact, there are many experts who would argue that Moore's law has been dead for quite some time. For one thing, while the GHz clock speeds we take for granted in our devices are certainly hugely impressive compared to the kHz bandwidth of ENIAC, there has been very little improvement in clock speed over the past decade—the industry has hit a wall. That's because cramming billions of nanotransistors onto a chip whose active area is sometimes less than 1 mm^2, and then switching them billions of times a second, generates lots of thermal energy that has to be removed somehow. Moreover, the faster the transistors switch, the more power is required. In effect, clock speed is a victim of the heat death of nanoelectronics.

Ironically, although ingenious nanotechnological innovations have so often been the saviour of the semiconductor industry, keeping Moore's law viable, it is nanoscale physics that will also ultimately be its nemesis (along with cold, hard economics.) The heat death issue, and the associated economic challenges to push silicon to ever-higher bandwidths, are just one aspect of the problem. Today's three-nanometre feature size is less than 10 silicon atoms across. We are firmly in the quantum realm at this length scale and the wavelike character of electrons therefore plays a key role in the operation of the device. It is the quantum nature of matter at the nanoscale that presents a major barrier to pushing Moore's law down to ever-smaller feature sizes. To understand why, we need to take a moment to consider the operating principle at the core of CMOS technology: transistor switching.

Breaking down the barriers: tunnelling electrons

Transistors are essentially a three-terminal switch that, in CMOS architecture, comprises a source, a drain, and a gate electrode. By applying a voltage to the gate electrode the flow of electrons—that is the electric current—between the source and drain can be switched on or off, placing the transistor in a '0' or '1' state. This is the basis of the binary logic that drives so much of our technology. When the gate length is at the micron, hundreds of nanometre, or even tens of nanometre length scale, the transistor can be controllably, and rapidly, switched on and off. We are, in effect, in the classical limit—the quantum nature of electrons does not have a huge influence on device operation.

Shrinking the gate length to sub-10 nm dimensions, however, changes the nanoscale physics of the situation dramatically. On these scales, quantum mechanical tunnelling—exactly the same tunnelling effect that is exploited in the scanning tunnelling microscope—becomes of key importance and can dominate the operation of the transistor. Electrons no longer 'see' the barrier presented by the insulating silicon oxide. Instead, they tunnel straight through it because their wavefunction penetrates the barrier and, in essence, shorts out the oxide between the source and the drain. This creates a leakage current and makes the transistor exceptionally unreliable because even in its nominal 'off' state, there can still be spontaneous current flow due to tunnelling.

A key aspect of tunnelling is that the process is extremely sensitive to the barrier width—it's this sensitivity that lends the STM its exceptionally high resolution (because very small changes in tip-sample separation lead to very large changes in the measured tunnel current). But what's a boon for the STM is a bane for CMOS nanoelectronics: as feature size gets smaller and smaller, the probability of electrons tunnelling through the oxide goes up exponentially. Each time the barrier width reduces by the diameter of just one atom, the tunnelling probability increases by a factor of hundreds. Although current research is focused on what are called high k dielectrics—replacements for silicon oxide that have insulating properties selected to provide less penetrable barriers—in many ways this is simply postponing the inevitable. Moore's law can't continue indefinitely because quantum tunnelling is ultimately going to short out the devices.

If we can't engineer around quantum physics, why don't we just embrace it instead? This is exactly the principle at the core of a quantum computer, in which the wavelike characteristics of matter, including tunnelling, form the basis of an entirely new approach to computing and device technology. However, and contrary to the more excitable pundits and predictions out there, quantum computing is not going to completely replace conventional computer technology. Sure, a quantum computer can solve problems and run algorithms that a classical computer would find impossible, or that would take an impossibly long time to complete (millions or billions of years). But the converse is equally true: classical computers are already better than their potential quantum counterparts at very many tasks, including, as just a few everyday examples, email servers, word processing and spreadsheets, MP3 and MP4 players, and graphic design applications. Quantum computers won't supersede classical computing; they'll complement it.

In any case, and although still a very long way from commercial viability, silicon nanolithography was pushed all the way down to the single atom limit almost 30 years ago. Joseph Lyding and colleagues at the University of Illinois at Urbana-Champaign, in collaboration with Phaedon Avouris and Robert Walkup at the IBM Thomas J. Watson Research Center at Yorktown Heights in New York, showed back in 1995 that by injecting tunnelling electrons (from an STM tip) into a silicon surface covered with a single atomic layer of hydrogen, it was possible to break the Si–H chemical bond and desorb H atoms one at a time. An STM tip parked over a hydrogen atom injects enough energy via tunnelling electrons to shake up the Si–H bond (literally) until eventually the hydrogen atom escapes the potential well in which it was hitherto trapped. This is the ultimate in top-down nanotechnology and nanolithography: extraction of single atoms.

Examples of single atom silicon nanolithography are shown in Figure 9. Unlike the commercial processes at the heart of CMOS technology, this type of atomic precision patterning is, at the moment at least, restricted to a very specialized, ultrahigh vacuum environment—single atom devices are unlikely to make it into a smartphone any time soon. This is because the removal of an H atom from the hydrogen-terminated surface produces what is known as a dangling bond on the now 'decapped' silicon (see schematic in Figure 9(a)). If taken outside the ultrahigh vacuum environment that silicon atom (and others that have similarly had their capping hydrogen removed) will react very quickly indeed. So, for now, atomically precise lithography of this type is a tool for basic research, where the limits of our control of matter are explored in very controlled conditions that are challenging, if not

9. **Atomically precise nanolithography.** (a) Schematic of STM-induced single atom removal. The STM tip is positioned above a single hydrogen atom (small filled circle) that is bonded to a silicon atom (open circle.) On this particular surface, the silicon atoms pair up to form rows of dimers (i.e. pairs); each dimer atom is initially capped with a hydrogen atom. The STM tip injects a flow of tunnelling electrons that excites the Si–H bond and causes the hydrogen to desorb. (b) One of the first examples of atomic precision lithography of this type, pioneered by Joseph Lyding and colleagues at the University of Illinois at Urbana-Champaign. The bright features are rows of silicon dimers that have had their capping hydrogen removed by the STM. The arrows point to areas where both hydrogen atoms of a dimer have desorbed. (c) A single atom transistor fabricated using STM-induced hydrogen desorption. The small dot at the centre of the image, between the two brighter contact regions, is the active area of the transistor and is just a single atom wide.

practically and economically impossible, to transfer to a manufacturing process.

In the next chapter we'll take a look at a number of next-generation approaches to information processing that are enabled by the type of atomically precise nanolithography shown in Figure 9 and, more generally, via the control of single atoms and molecules. But scaling this type of atom-by-atom or molecule-by-molecule approach up to a commercially viable technology or manufacturing process presents an immense practical challenge because, in essence, it's *serial*. Photolithography has been the bedrock of the electronics and semiconductor industries for so long precisely because it's a technique that patterns a chip not one component at a time but all at once: it's a parallel process. Similarly, to structure matter at the nanoscale, we don't have to rely on scanning probes to shunt atoms and/or molecules around one at a time. Nature continually exploits the interatomic and intermolecular forces described in Chapter 1 to structure matter at the nanometre level (and far beyond), in parallel. Instead of artificially chiselling away at matter to define nanostructures, the natural world adopts a polar opposite approach: it builds from the bottom up.

Self-assembling nanostructures

On the left of Figure 10 is an electron microscope (EM) image of a set of 7 nm transistors created using the type of advanced semiconductor patterning described in the previous section; nanoscale structural order has been imposed, top down, with a remarkably high level of precision. By its side, an equally

10. (a) Electron microscope image of 7 nm field effect transistors (FETs) fabricated on a silicon chip. (b) Self-assembled lattice of proteins (an S-layer) from the *Sulfolobus* archaeon. (Archaea are single-celled organisms that are similar to, and for some time were classified as, bacteria.) The inter-protein separation is approximately 20 nm.

impressive example of nanopatterning—an EM image of a protective lattice of proteins of a type that encages many Bacteria and Archaea. S-layers, as they are known, comprise proteins that are among those most commonly found on Earth and not only define the shape of the organism but also play a central role in many of its essential biological and biochemical functions: protection, adhesion, and cell division being key amongst them. And yet the S-layer lattice, with its intricate nanoscale organization rivalling that of the state-of-the-art silicon chip, didn't require a £2 billion semiconductor foundry for its fabrication. Instead, it formed spontaneously, without any need for external technological assistance. In other words, it *self-assembled*.

Nature is a wonderfully inventive nanotechnologist. Millennia before humanity had gained control over the nanoscopic world, nature was exploiting—and continues to exploit—interatomic and intermolecular forces to structure matter on the smallest scales, building a bewildering array of sophisticated nanostructured patterns in which, in each case, form follows function. In their overview of the contributions to a self-organization-themed issue of the *Philosophical Transactions of the Royal Society*, Roland Wedlich-Söldner and Timo Betz of the Cells in Motion initiative at the Westfälische Wilhelms-Universität Münster in Germany described self-organization as the fundament of cell biology:

> Ultimately, self-organization lies at the heart of the robustness and adaptability found in cellular and organismal organization, and hence constitutes a fundamental basis for natural selection and evolution.

You may have noticed that I switched from using the term *self-assembly* to referencing Wedlich-Söldner and Betz's description of *self-organization*. Some scientists would treat these two terms as effectively synonymous. Others, including myself, draw a particular distinction: self-assembly occurs when the molecular system is close to thermodynamic equilibrium;

self-organization is what happens when there is substantial energy or matter flow, that is the system is far from equilibrium. Thermodynamic equilibrium simply means that any part of a system looks much like any other. Think of opening a bottle of particularly fragrant perfume in a room. At the instant the bottle is opened, the system—which in this case can be thought of as the air in the room—is tipped away from equilibrium; there is a localized 'injection' of matter (and/or energy) at a certain point, which means that the distribution of molecules in the room is not uniform. Wait a little while, however, and the system will *equilibrate*, that is the perfume molecules will diffuse across the room—due to collisions with their neighbours—until they are distributed uniformly. Equilibrium is restored.

In a system that self-assembles, the molecules are largely free to interact solely according to the type of intermolecular potential described in Chapter 1. Generally, the forces underpinning molecular self-assembly involve interactions whose bond strength is much lower than the traditional covalent and/or ionic bonds that hold the atoms in inorganic crystals such as silicon, copper, and gold together. In addition to the van der Waals forces we've already discussed, self-assembly generally involves other types of weaker interaction such as: hydrogen bonding (responsible for the rather bizarre behaviour of water, and whose explanation has a long and chequered history); hydrophobicity and hydrophilicity (which again necessitate an understanding of the interactions of water, and which arise fundamentally from the entropy of mixing two different substances—there's more on entropy below); and coordination chemistry (which is rather like covalent bonding except that both electrons involved in the bond come from the same atom or molecule).

Self-assembly of well-ordered structures such as the S-layer shown in Figure 10 requires intermolecular forces that are relatively weak because, otherwise, the molecules will get locked in place. If the bonding is too strong, there is no flexibility in the assembly process—no way to correct errors. It's a little like building a structure with LEGO blocks that, once fixed in place, can never be detached. Every error in block placement is irreversible. However, if the LEGO blocks can be readily detached then they can be moved into the correct site. But, unlike LEGO, in molecular self-assembly there's no sentient force guiding each and every molecule into place. Instead, thermal energy drives the molecules along essentially random trajectories—the same effect underpinning Brownian motion—so that they can explore the energy landscape, eventually finding their preferred binding site. If, instead, a molecule finds itself bound in a less favourable site, as long as the binding energy is not too strong (as compared to the thermal fluctuations) it will at some point be released and will diffuse to a more energetically beneficial position.

It's not just the intermolecular binding energy that controls the molecular ordering, however. Entropy not only plays a key role but can also sometimes *dominate* the assembly process. As briefly noted in the previous chapter, traditional explanations of entropy generally describe the phenomenon in the context of ever-increasing disorder but, at best, this doesn't capture the core principles underlying the science, and, at worst, is entirely misleading. Frank Lambert (1918–2018), an American academic who tirelessly advocated for changes in the teaching of entropy in undergraduate courses, put it memorably in a title of a paper he published in 1999: 'Shuffled cards, messy desks, and disorderly dorm rooms—examples of entropy increase? Nonsense!' Entropy instead relates to the configurations of the components of a system, quantifies their relative probabilities of occurrence, and accounts for just how energy is distributed (or dispersed) across those configurations. As such, entropy is closely linked with the mathematics of permutations, combinations, and probabilities; it is essentially statistical in nature. In molecular self-assembly the various configurations in question relate to the possible rotations, vibrations, and translations of the molecules.

A particularly compelling and beautiful example of the influence of entropy on molecular self-assembly comes from the work of my colleague Peter Beton, and his fellow researchers in the University of Nottingham Schools of Physics & Astronomy, and Chemistry. Take a look at Figure 11. Each of the 'rods' in the STM image represents the backbone of the planar organic molecule whose structure is also shown in the figure. (The molecule is p-terphenyl-3,5,3,́5´-tetracarboxylic acid, or TPTC for short, but its name isn't important. All that matters is that the three carbon rings that form the 'backbone' of the molecule appear bright in an STM image.) At first glance, the arrangement of molecules appears highly ordered, with each molecule seemingly adopting the same configuration with respect to its neighbours. But take a closer look. Try, for example, following a row of molecules across the image. You will find many examples of where the orientation

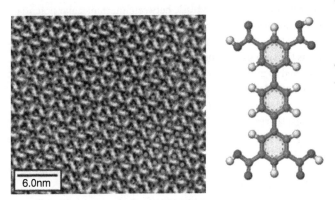

11. On the left, a scanning tunnelling microscope image of a single self-assembled layer of the molecule whose chemical structure is shown on the right. This is p-terphenyl-3,5,3,́5´-tetracarboxylic acid; the three carbon rings that form the central structure of the molecule make the dominant contribution to the STM signal. Each of the 'rods' seen in the image on the left arises from the three-ring backbone of the molecule. The arrangement of molecules in the lattice is fundamentally governed by entropy.

of a molecule is 'out of sync' and breaks the pattern; the molecular organization is much less ordered than it first appears.

It turns out that it's not the variation in the intermolecular potential that plays the key role in defining self-organization in this case (because there is little difference in the molecule–molecule binding energy for each of the molecular configurations). Instead, the pattern arises almost entirely from the contribution of entropy—in other words, it is the number of possible ways of organizing ('tiling') the molecules on the surface that determines the optimal molecular lattice. The particular type of molecular organization seen in Figure 10 is, in essence, much more probable than competing configurations, just as it is overwhelmingly more probable for those perfume molecules to diffuse across the room than to stay together.

There is always a constant tussle between intermolecular potential energy (as described by the types of interactions highlighted in Chapter 1), and entropy, that is just how the energy is distributed across the various configurations of the system. This is not restricted to nanoscience and nanotechnology. Right across the spectrum of length scales, from atoms to galaxies, the balance of interaction energies and entropy determines how any system evolves; thermodynamics focuses not just on energy but on *free* energy, a quantity that accounts for the balance between intermolecular potential, the influence of the environment, and entropy. At the nanoscale we're concerned with controlling the basic building blocks of matter—atoms, molecules, nanoparticles—and steering them to form the type of structures we need. Nanotechnologists therefore spend a great deal of time attempting to find the 'sweet spot' that balances the contributions of intermolecular potential, entropy, and external influences and stimuli (heat, pressure, light . . .) in just the right way.

Sharon Glotzer, Professor of Chemical Engineering at the University of Michigan, has led an innovative, inspiring, and highly influential programme of research that has exploited entropy to drive *more*, not less, order. As Glotzer put it in an interview for *Quanta* magazine,

> What happens is the particles try to maximize the amount of space that they have to wiggle around in. If you can wiggle, you can rearrange your position and orientation. The more positions, the more options, and thus the more entropy . . . what these systems want to do is space out the particles enough so that it maximizes the amount of wiggle room available to all the particles. Depending on the particle shape, that can lead to extremely complicated arrangements.

Glotzer and her team focus on what is known as *emergence*—how simple objects, following very simple rules, can produce surprisingly complicated, collective behaviour. (Starling murmuration is a very good real world example of this type of collective interaction, although starlings are perhaps not the simplest object one could imagine.) They study how (nano)particles of different shapes and sizes self-assemble and self-organize. Given the core role of entropy in driving the formation of ordered structures, Glotzer firmly believes that we should extend the definition of intermolecular and inter-particle interactions to incorporate entropy in a much more up-front manner: the entropic bond.

Out of equilibrium

In an ideal world, self-assembly would continue until each and every atom, molecule, and/or nanoparticle was in its most favourable, lowest energy state. This is the most stable configuration of and is often the ultimate goal of many nanotechnologists who exploit self-assembly to generate particular nanoscale patterns. But little is ideal in science (despite the extensive use of idealizations in physics, for one). Hence, self-assembly can often produce structures that are not the most stable state but are what is known as *metastable*. A metastable state refers to a system that has not reached its lowest possible energy—the molecules (or atoms, or nanoparticles) have got locked into a configuration from which it could take a very long time to escape. This time period could range from anywhere from microseconds to millions of years (and beyond) but the process can generally be accelerated by injecting thermal energy, that is by heating things up. A good example of a metastable structure is diamond. The most thermodynamically favourable state for the carbon atoms to adopt is not the beautiful crystalline form of the diamond ring but the much less aesthetically appealing graphite (i.e. pencil 'lead'). Given enough time (although it will be a very long wait indeed), diamond will fade to black.

Colloidal nanoparticles are particularly prone to getting trapped in far-from-equilibrium states. To fabricate solid state devices that exploit the electronic and optical properties of nanoparticles, we need those particles in the solid state. That means we must transfer the nanoparticles from their suspended state in the solvent to a solid substrate, on which we can fabricate electrical connections so as to measure and control electron flow through the nanoparticle assembly. There are many parallels here with the physics of coffee stains: we have a solute (the nanoparticles) suspended in a solvent. Place a droplet of that suspension on a surface and let it dry. What will happen?

It turns out that a lot of rich and fascinating physics, and physical chemistry, underpins what at first appears to be a very simple experiment. If the solvent evaporates quickly then the nanoparticles don't have enough time to diffuse to their equilibrium state; they're left high and dry because, without their surrounding solvent, they can no longer move. Slow the rate of solvent evaporation, however, and the particles can explore the energy landscape much more comprehensively, finding more favourable configurations as they diffuse from position to position in the growing nanoparticle network. The nanoparticles are, as a colleague once put it, passengers on the tide of the solvent—they act as tracers for just how the solvent evaporates. A detailed study of this type of drying-mediated self-organization, by Eran Rabani and co-workers (at Tel-Aviv University, MIT, Harvard, and Columbia University, respectively), provided key insights into just how inter-particle, solvent–article, and solvent–solvent interactions combine to produce a remarkable array of self-organized structures (see Figure 12).

One especially prevalent type of pattern in not just nanoparticle systems but any self-assembly or self-organization process involving deposition from solution—including polymers, proteins, DNA, and, more generally, molecules of all types—is the foam, or cellular network (Figure 13). Holes nucleate due to evaporation and subsequently expand in the solvent film, with the nanoscopic 'cargo' tracking back as the void in the liquid grows. The final state of this process is a self-organized state which can be structured on a number of length scales (Figure 13) and is best described as a foam or cellular network, where 'cellular' is used in the geometric, rather than biological, sense. (There are key parallels with the physics and physical chemistry of coffee stains, although the patterns formed by the nanoparticles are rather more complex.)

12. **Examples of far-from-equilibrium organization of gold nanoparticles on a silicon wafer. Each atomic force microscope image shows the organization of the nanoparticles on a scale of microns (or tens of microns) rather than nanometres—the individual nanoparticles are not observed at this magnification level. Instead, the patterns arise from the collective response of large numbers of particles as the solvent in which they are carried flows and evaporates.**

If we can control just how the solvent evaporates, we can influence the final destinations of the nanoparticles through both self-assembly and self-organization. A number of groups worldwide, including our own at Nottingham, have done just that. Moreover, patterns like these are not confined to nanoparticles and are observed in a wide variety of systems spanning length scales ranging from nanometres to kilometres (and far beyond).

Often there is a focus on attaining high degrees of molecular or particle order—nanotechnologists tend to aim to produce the most flawless crystalline states possible, with each molecule aligned with its neighbours. But there is beauty, and function, in messiness and disorder too. Indeed, the neural network that is firing inside your skull as you read this is very far from a perfectly ordered or highly symmetric state of matter; it's the connectivity that's key, not the symmetry. Similarly, although nature produces crystalline lattices that are exquisitely ordered (silicon is a very good example), it also generates much less ordered, but no less functional, cellular and network structures across a quite remarkable range of length scales, including at the nanometre level. Figure 13 shows just a few examples of the ubiquity of foams and cellular networks across nature, including the striking self-organized micro- and nanoscale patterns formed by diatoms—unicellular microalgae that form a major part of the ecosystems around us, generating up to 50 per cent of the oxygen produced on Earth each year, and comprising nearly half of the organic material and organisms found in the oceans. (It's worth noting that life itself is a far-from-equilibrium state; if you'll excuse the morbid thought, our ground state is death.)

We can both mimic and steal from nature's ability to structure matter across multiple length scales in this way—strategies that have been described as biomimicry and, with tongue slightly in cheek, biokleptocracy, respectively. If we're going to steal from nature's nanotechnology then one might reasonably imagine that our primary choice of biomolecular system to 'rip off' would be

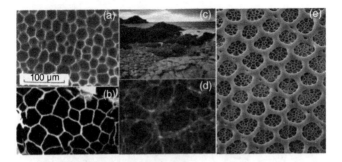

13. **Foams and cellular networks in nature. (a) Cross-section of a cork from a wine bottle; (b) the hide of a giraffe; (c) the Giant's Causeway in Antrim, Northern Ireland; (d) simulation of the large-scale structure of the universe—the arrangement of galaxies is best described as a cellular network (or 'the cosmic foam'); (e) the intricate micro- and nanostructured pattern formed by the unicellular alga known as the diatom (cf. the first frame of Figure 12). The foam-within-foam motif is common in nanostructured matter.**

DNA. With an extraordinary propensity for biological information storage (driven by molecular recognition due to base pairing), deoxyribonucleic acid is an exceptionally powerful platform for the bottom-up generation of intricate nanoscale structures whose *programmable* complexity can easily match that of the advanced silicon device architectures described in the previous sections. There will be more on DNA nanotech in a later chapter, but, for now, I'll briefly highlight the work of Paul Rothemund, of the California Institute of Technology (Caltech), who in 2006 pioneered the technique known as DNA origami (building on previous work by Erik Winfree (also at Caltech) and Nadrian Seeman (New York University) on DNA nanotechnology). Rothemund's approach involves folding long, single-stranded DNA molecules into arbitrary shapes. At first, these were two-dimensional objects and patterns but DNA origami now involves the design, encoding, and subsequent 'steered' self-assembly of sophisticated three-dimensional shapes, including rods, spheres, cubes, and much more complicated objects such as nanoflasks and gears (Figure 14).

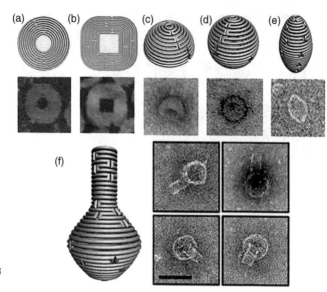

14. **DNA origami.** DNA can be chemically programmed to assemble nanostructured objects and patterns in both two and three dimensions. The examples shown here include (a), (b) rudimentary 2D patterns and their corresponding atomic force microscope images (although DNA origami has also been used to fabricate much more sophisticated 2D patterns including snowflakes, stars, smiley faces, words, and maps); (c), (d), (e) 3D objects—hemisphere, sphere, and ellipsoid, with corresponding transmission electron microscope (TEM) images, and (f) a flask, again with TEM images taken from various angles.

The carbon flatlands

We opened this chapter with a discussion of the history of silicon nanotechnology. We're going to close with a consideration of a very close neighbour of Si in the periodic table: carbon. The two elements are in the same group of the table, each having four valence electrons, and so on first consideration they might be thought to be chemically rather similar. Yet from many perspectives they could not be more different. Organic life is carbon based; organic chemistry is, essentially, the chemistry of carbon. But our technology—including, in particular, the nanoelectronics systems that power so much of our information age—has traditionally been inorganic. That's changing (inexorably slowly, admittedly), and is driven, not least, by the pioneering advances in molecular self-assembly and DNA nanotechnology outlined above. Molecules typically contain a variety of elements, however—the biochemistry of DNA, for one, relies on phosphorus, nitrogen, oxygen, and hydrogen alongside carbon. Is a nanodevice based on crystalline carbon alone possible?

Enter the *wunderkind* material known as graphene (Figure 15). Just a single atom thick, yet the strongest material ever discovered, graphene is not just a new crystalline form of pure carbon—an *allotrope*—but it's a two-dimensional solid. Edwin Abbott's 1884 novel *Flatland: A Romance of Many Dimensions* is both a scathing satire of Victorian society and a clever investigation of reduced dimensionality. Abbott's fictional flatland—if not its inhabitants—was, in essence, experimentally realized with the discovery of graphene via what is technically described as micromechanical exfoliation but which is much better known as the sticky tape method. As Kostya Novoselov—who, along with his colleague Andre Geim at the University of Manchester, was awarded the 2010 Nobel Prize in Physics for the discovery of graphene—describes in his Nobel Prize acceptance speech, it's possible to 'peel off' layers from a graphite crystal (and subsequently transfer them onto another surface) using Sellotape, all the way down to a pure, single-atom-thick 2D lattice. Indeed, if you've ever used a pencil, you'll have carried out a somewhat similar process of exfoliation of graphite (i.e. the pencil 'lead') to make your mark. Although the original isolation of, and experiments on, graphene used the sticky tape approach, for scale-up to the throughput levels required for viable manufacturing of devices, various types of direct growth of

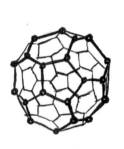

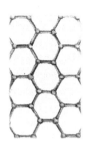

Buckminsterfullerene (C_{60}) Nanotube Graphene

15. Carbon nanotechnology. Carbon forms nanostructures that confine electrons in all three dimensions (the buckminsterfullerene molecule, C_{60}), in two dimensions (graphene), and in one dimension—the carbon nanotube.

graphene on substrates, including chemical vapour deposition and molecular beam epitaxy, have been developed and are now routinely exploited.

While nanoparticles confine electrons in all three dimensions, graphene imposes confinement in a 2D plane. Prior to the discovery of graphene, two other pure carbon nanosystems generated comparable levels of excitement among nanoscientists: buckminsterfullerene or C_{60}, and carbon nanotubes. Buckminsterfullerene, or 'buckyball' for short, is almost perfectly spherical—the most symmetric molecule in nature (technically of icosahedral symmetry)—comprising 60 atoms organized in a manner that is identical to the stitching on a traditional football: each carbon is at a vertex between 20 hexagons and 12 pentagons (as shown in Figure 15). In other words, it's a carbon nanoparticle, and, just like the metal and semiconductor nanoparticles we discussed earlier, the electrons in C_{60} are confined in all three dimensions. Moreover, the diameter of the buckyball is just a little under 1 nm—in many ways it's a molecule almost custom-built for nanoscience, and it has played a central and revolutionary role in the development of many aspects of nanoscience since its discovery (by Harry Kroto, Richard Smalley, and colleagues, for which they subsequently won the Nobel Prize in Chemistry (in 1996)).

Carbon nanotubes, on the other hand, are nanoscopic cylinders of pure carbon: essentially a graphene sheet rolled up and 'zipped up' along an edge. In a nanotube, electrons are confined so that they are free only in one dimension—along the tube. In each case—buckyballs, nanotubes, and graphene—the dimensionality of the system is key in defining the nanoscale properties of the confined electrons. In graphene in particular, the limited dimensionality, coupled with the honeycomb arrangement of carbon atoms, gives rise to some truly unique electronic properties. For one, their interaction with the honeycomb lattice, and the rather unusual structure of the lattice itself, mean that the electrons responsible for carrying electric current in graphene behave more like photons: they act as if they have zero mass. Moreover, they move as if the speed of light were 10^6 m/s (rather than its actual value of 3×10^8 m/s in vacuum), opening up the possibility of doing bench-top particle physics experiments rather than requiring multi-billion-pound accelerators. Particle physicists, who have been sometimes known to refer to their condensed matter/nanophysics colleagues as 'squalid state' scientists, have been paying particular attention to graphene.

Graphene is not, however, just a playground for novel fundamental physics; its unique properties mean that it has an exceptionally wide range of applications. These run the gamut from transparent conducting coatings for solar cells and touch screens, sensitive gas detectors and nanoscale 'noses' (i.e. molecular detection), ultrafast detectors of light, gas barriers, strain gauges, and water filtration and sanitization. And nanoelectronic devices that might possibly supersede silicon CMOS. On this latter application, however, graphene still has quite some way to go before it can be a serious challenger to silicon. Some of these hurdles are due to fundamental physics—for one, in its native state graphene doesn't have the energy band gap required for a variety of nanoelectronic applications (although the required gap can be induced via careful chemistry and by deliberately distorting the carbon lattice).

Even when the physics and electronics engineering issues are surmounted, however, there remains the hard economic reality: any move away from industry-standard silicon-based infrastructure to incorporate graphene, or any of the various other similar 2D materials systems that have been exploited in its wake, will require a radical rethinking of design and fabrication strategies. That will be costly. But as we'll explore in the next chapter, graphene is not the only route towards revolutionary electronics and information processing at the nanoscale.

Chapter 4
It from bit, bit from it

Ada Lovelace was born Augusta Ada Byron, daughter of Lady and Lord Byron, in 1815. Due to her mother's mathematical training, Ada was tutored in mathematics from an early age—a very unusual education for a woman of the time—and demonstrated a prodigious ability and appetite for the subject. After being introduced to Charles Babbage at a party in 1833, Lovelace became fascinated by the workings of Babbage's difference engine, a mechanical computer that used a sophisticated set of gears to calculate the solutions to mathematical and arithmetical problems. In translating from the original French of an article on the difference engine, Lovelace added extensive notes of her own, including, in particular, a description of a set of steps that could be implemented on the machine in order to solve specific mathematical problems. These notes are widely recognized as the first example of a complex algorithm for a machine (although opinion is divided as to whether Lovelace could be said to have written the first computer program).

Lovelace understood the deep and subtle links between matter and mathematics, and between information and the material world, more than a century before those connections were appreciated by the scientific community: 'In enabling mechanisms to combine together general symbols in successions of unlimited variety and extent, a uniting link is established between the operations of matter and the abstract mental processes of the most abstract branch of mathematical science.' In the 1970s, physicist John Archibald Wheeler also recognized this essential interplay of information and matter, coining the pithy and poetic 'It from bit' aphorism. We encode and manipulate data through physical processes, be they the basis of silicon processors, DNA, neural pathways, or, as we'll see, interacting atoms. (Note that each of these, not coincidentally, involves nanoscale processes and processing.) For Wheeler, information is much more fundamental than energy, matter, fields, and forces, that is the foundational framework of physics: 'Otherwise stated, every physical quantity, every it, derives its ultimate significance from bits, binary yes-or-no indications, a conclusion which we epitomize in the phrase, it from bit.'

Nanotechnology has progressed to the point where we can represent binary information on sub-nanometre scales, and where information storage can involve writing data with a resolution much better than that defined even by the atomic limit; it's possible to store information on a scale considerably smaller than the size of an atom. We're far beyond nanotechnology at that stage and into the realm of picotechnology—the manipulation of matter on sub-atomic length scales. Lovelace's far-reaching insights into the connections between matter, mathematics, and the manipulation of information are especially pertinent in this context, as we'll see repeatedly throughout this chapter.

DNA computing

Before we get to the atomic limit, we need to revisit the DNA-driven nanofabrication covered in the previous chapter. Self-assembly, be it driven by the type of molecular recognition that underpins the DNA origami technique or not, is a powerful method for generating a variety of nanoscale (and microscale) patterns and structures. But those patterns are much more than pretty pictures. Just as Lovelace highlighted, the links between matter and mathematics are legion: we can *compute* with nanostructured patterns.

As we've seen, DNA can be programmed to produce different forms of periodic and aperiodic pattern. Those output patterns can, however, also be the result of a computation: logical operations—ANDs, NANDs, NOTs, and NORs, for example—can be encoded into the intermolecular interactions between different DNA molecules and the final output 'read out' simply by imaging the resulting self-assembled pattern. Or, as demonstrated by Leonard Adleman of the University of South California in 1994, much more sophisticated problems beyond binary logic can be encoded in DNA. In the first published demonstration of DNA computing, Adleman showed that the classic travelling salesman problem—'Given a list of cities and the distances between each pair of cities, what is the shortest possible route that visits each city exactly once and returns to the origin city?'—could be solved via DNA nanotechnology.

In the decades since that pioneering demonstration, Erik Winfree's group at the California Institute of Technology (Caltech), in particular, has built on Adleman's work to drive DNA computation to impressively high levels of sophistication and control, including in the implementation of DNA nanotubes to yield a *reprogrammable* self-assembly system. This is nanotechnology at its most powerful and elegant: a convergence of traditional disciplines—(bio)chemistry, biology, physics, engineering, computing, and more—at the nanometre scale, with nanostructured matter serving to embed information and computation.

To a physicist, however, DNA is an exceptionally large, complicated molecule. We are reductionist to our core, preferring to reduce the world and the universe to its simplest possible building blocks. What are the limits of miniaturization when it comes to computing? How small can we go? Feynman mused over exactly this question not only in 'There's Plenty of Room...' but also in a course he gave at Caltech from 1983 until 1986 entitled 'Potentialities and Limitations of Computing Machines'. There has been substantial progress in atomic and nanoscale computing since Feynman gave that course.

Atomic bits

What's the simplest possible electronic component you can imagine? Some might say a resistor; something as rudimentary as a length of wire has a finite resistance. But that's not a particularly functional choice. The simplest component that fulfils a controllable function is the switch. Its state can be flipped between on and off, or, in terms of binary choices, 0 and 1, and that provides a method of not only encoding information but also controlling states and processes. Switches are thus everywhere around us; there's a humble, but extremely helpful, power switch on every household appliance.

Shortly after Don Eigler and colleagues controllably positioned atoms for the first time to spell out the IBM logo, they demonstrated an atomic switch due to the transfer of a single Xe atom between the tip of an STM and a nickel surface. The switch was flipped, that is the Xe was transferred either from the surface to the tip or vice versa, via the application of a voltage pulse. Earlier the same year (1991), In-Whan Lyo and Phaedon Avouris at IBM's Yorktown Heights research labs had demonstrated controlled extraction of a single atom from a silicon surface. The IBM researchers had previously demonstrated an important device characteristic known as negative differential resistance—by which, due to a phenomenon known as resonant tunnelling, the current between two electrodes can *drop* as the voltage is increased (completely contrary to Ohm's law)—for a single atom.

Those early pioneering experiments by the IBM Almaden and IBM Yorktown Heights teams stimulated a flurry of interest in using

STM and AFM to implement electronic device functionalities and computing principles at the atomic, molecular, and nanometre scales. In this *Very Short Introduction* I can't begin to do justice to that body of work, which is due to the efforts of a large number of research groups across the world and spans very many decades. Instead, I'll focus on a few influential highlights.

In yet another ingenious experiment, Eigler's team used what they called molecular cascades to encode Boolean logic. They lined up CO molecules in just the right way so that when the first molecule in the row was moved (with an STM tip), it triggered a chain reaction (Figure 16). Like a nanoscopic game of dominoes, each molecule interacted with its neighbour in sequence, changing its configuration, all the way to the end of the line. By careful design of the molecular 'tracks', which enabled precise control of the cascade of intermolecular interactions, Eigler's team encoded Boolean logic (an AND gate), and a variety of other digital functionalities. This type of molecular logic is exceptionally slow by comparison with silicon CMOS and other electronic architectures, but the goal here was not to fabricate a device that could replace conventional gadgets. Instead, the molecular cascade strategy implements logic in an entirely novel manner and we can learn much about the fundamentals of computing by playing with matter at the nanoscale in this way.

I've thus far avoided too many references to Feynman's celebrated 1959 'There's plenty of room at the bottom' speech, not only because it's cited in just about any article on nanotechnology but also because it turns out it was rather less influential on nanotechnology pioneers than initially assumed. Although Feynman is perceived to have had a massive influence on the origins and evolution of the science of the ultrasmall, like so many other scientific developments, the truth of the matter is rather more 'non-linear' and convoluted. In an article for *Chemistry World* written in 2009 to celebrate the 50th anniversary of 'There's plenty of room at the bottom', the science writer Philip

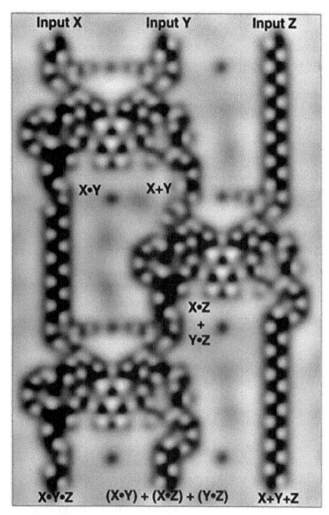

16. **Molecular computing.** A variety of logic operations (where the +
symbol represents a logical OR and the • represents a logical AND)
implemented via molecular cascades triggered by an STM tip.

Ball examined its legacy and influence. Ball highlights an analysis by Chris Toumey, an anthropologist at the University of South Carolina, which revealed that in the two decades following the publication of the transcript of his talk, Feynman's legendary lecture attracted a grand total of *seven* citations—and one of these was far from complimentary: 'completely vacuous as far as the real world is concerned'.

Remarkably, the inventors of the STM, who did more than anyone else to realize Feynman's dream of atomically precise engineering, were entirely unaware of '...room at the bottom'. Nonetheless, and due in no small part to all of the publicity, Feynman's speech has certainly been read and reread by many of the current generation of nanoscientists. In that sense it continues to be a major influence, this passage in particular:

> But I am not afraid to consider the final question as to whether, ultimately—in the great future—we can arrange the atoms the way we want; the very atoms, all the way down! What would happen if we could arrange the atoms one by one the way we want them (within reason, of course; you can't put them so that they are chemically unstable, for example)?

Feynman, who died in 1988, would have been overjoyed to see those words from his now-iconic speech translated into atoms in a *tour de force* demonstration of atomic encoding in 2016.

In Figure 17, each dark dot is a single atom hole in a lattice of chlorine on a copper surface. The researchers who encoded that passage from Feynman's speech—namely, Floris Kalff and co-workers at the Kavli Institute of Nanoscience in The Netherlands—used the STM tip to rearrange the vacancies according to a binary ASCII code: each bit comprises a vacancy/ hole ('0') and a Cl atom ('1'). When complemented by a number of marker patterns to define the start and end of lines, and to identify blocks where defects prohibited the manipulation of

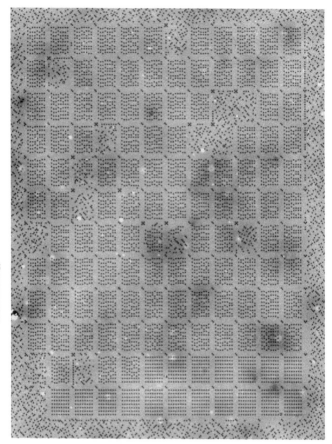

17. Plenty of room at the bottom: Feynman's celebrated 1959 speech encoded in single atom vacancies in a chlorine lattice.

atoms, the researchers could build up intricate atomic arrangements of considerable complexity. The title of their paper, published in 2016, says it all: 'A kilobyte rewritable atomic memory'.

Impressive though the atomic kilobyte might be, it's certainly not the highest density data storage possible. In 2012, two alumni of

Eigler's IBM Almaden team, Hari Manoharan and Christopher Moon, led a team of researchers at Stanford to produce what is currently the world's smallest writing, as recognized by the Guinness Book of World Records. They wrote not in atoms but by controlling the electron density at a surface so as to encode information with sub-atomic precision, achieving information storage beyond the single atom limit (Figure 18). Manipulation of the electron waves formed at the surface of a copper crystal— exactly the same type of waves that give rise to the pattern within the circular quantum corral—produced the letters 'SU' as a modulation of the electron density. Because those free electrons can have a wavelength considerably shorter than the interatomic spacing at the surface, they can encode information in a smaller area or volume. There's more room at the bottom than even Feynman envisaged.

While these are all impressive examples of the control of matter at the atomic, molecular, and nanometre levels, there are a number of show-stoppers in terms of extending these approaches to a viable and scalable commercial technology. First, ultrahigh vacuum and cryogenic temperatures (~ 4 K) are hardly the most convenient of device environments. These limitations are exacerbated by the speed issue: although the molecular cascade is an extreme example, *all* scanning probe methods suffer from a severe lack of bandwidth. As we've seen, and until multiple tip SPMs capable of atomic resolution are developed, the technique is inherently serial and, therefore, exceptionally slow. But at a more fundamental level, a metal substrate is not at all a good platform on which to fabricate atomic or molecular scale devices. If we want to package the device and connect it to the outside world via electrical contacts, the metal substrate will simply short out the device.

This is just one of many reasons why semiconductors, and silicon in particular, have been the material of choice for solid state electronic and computing devices, at any scale. We've already seen

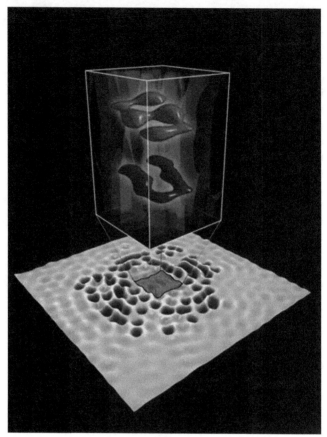

18. Sub-atomic precision: the smallest writing in the world. The letters SU have been encoded in variations in the density of electrons at a copper surface, with a precision smaller than the diameter of the atoms that comprise the surface.

that it's possible to pattern a silicon surface right down to the single atom level by desorbing hydrogen atoms with the STM tip. Although the use of STM is still a major hurdle in terms of upscaling this atomic lithography process to produce billions of devices on a single chip, the silicon platform means that the single atom devices are not shorted through the substrate and can be integrated with external circuitry, not just in principle but in practice. The ultimate limit in silicon-based information processing and logic at the single atom level involves the control of not just the charge state of silicon (see Figure 19), as demonstrated by Taleana Huff, Roshan Achal, and co-workers in Bob Wolkow's group at the University of Alberta, but, in essence, its magnetic properties. Given that silicon isn't a magnetic material, that may seem rather surprising.

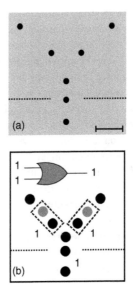

19. **Dangling bond logic. An OR logic gate fabricated on a hydrogen-terminated silicon surface by using the tip of an STM to remove individual H atoms. This produces what is known as a dangling bond on the underlying silicon atom. The charge state of each bond can be controlled by interactions with neighbouring dangling bonds, enabling logic operations to be carried out at the electron orbital level.**

Spins and songs

Despite its central role in so much of our technology, charge isn't the only property of an electron that we can exploit. Magnetism results from a very different property of electrons: their spin. (Technically, it's more correct to say that it's the interplay between charge and spin that underpins magnetic behaviour. But for reasons of space I'm going to have to skirt over some of the fine detail in the following. I thoroughly recommend Stephen Blundell's *Very Short Introduction to Magnetism* as a rather more detailed introduction to the fascinating subject of magnetism.)

What is spin? That's a very deep, multifaceted, and challenging question that ultimately has its resolution in a combination of relativity and quantum mechanics, and involves the Pauli exclusion principle at its core. Spin is a quantum mechanical property of not just electrons but all known elementary, sub-atomic particles (with the notable exception of the Higgs boson). The difficulty is that the electron—and this holds true for all other particles with spin—is not just a nanoscopic version of the spinning top with which we're familiar in our macroscopic, everyday world. Both systems possess angular momentum, and in this sense the electron is spinning. But an essential difference between the spinning top and the electron, other than scale, is that only for the latter is its intrinsic angular momentum quantized. It can take only two values—'up' or 'down'—and the alignment of a material's electrons in one or other of these directions defines its magnetic behaviour. Think of the electron as a tiny bar magnet that can be oriented with its north pole pointing up or down.

Physics students are often told that spin is a purely quantum mechanical effect with no classical analogy and that it's best just to think of spin only as a little arrow label, pointing upwards for an 'up spin', and downwards for a 'down spin'. Indeed, most professional physicists will tend to consider spin in this way. But this recourse to the traditional 'quantum mechanics is so very weird—it can't be understood in terms of our classical view of the world' argument greatly overstates the case when it comes to spin, and indeed many other quantum mechanical variables.

Classical physics is the limiting case of quantum physics: as length scales increase, or equivalently, the de Broglie wavelength approaches zero, the quantum evolves into the classical. There isn't a sharp, binary transition. Much like it is practically impossible to point to a rainbow and identify a well-defined point at which, say, red becomes orange, so too does quantum physics gradually transition to classical behaviour as the size of a system (and/or its interaction with its environment) increases. This is the all-important correspondence principle of quantum mechanics: the classical emerges from the quantum. Moreover, quantum mechanics is a theory for which observables—experimentally measurable quantities—are a defining feature of the framework. There is thus an intrinsic connection between the experiments we carry out on human length scales and the (sub-)nanoscopic world of the quantum.

When it comes to spin, two experiments over a century ago compellingly demonstrate this link between the macroscale and the nanoscale, or equivalently, between classical and quantum physics. Einstein and Dutch physicist Wander Johannes de Haas showed that changing the magnetic moment of a sample causes it to rotate. The magnetic moment arises from the total angular momentum of the electrons, including both their intrinsic spin and the momentum that results from the atomic orbital in which they're found. A change in quantum mechanical spin therefore manifests as a clearly observable change in the rotation of a body. Does this work in reverse? If we rotate a macroscopic sample can we change its magnetization? Yes, we can. Samuel Barnett showed that this is indeed the case as long ago as 1915, in a *Physical Review* paper with the laudably pithy title of 'Magnetization by rotation'.

I have taken this small detour into the physics of spin not only because of its role in cutting-edge nanotechnology (see below) but also to highlight once again that we should not think of the nanoscale, including quantum physics, as a weird domain that is somehow completely orthogonal to the everyday world with which we're more familiar. The nanoscopic connects with the macroscopic in a variety of unexpected ways.

Electronics relies on the control of electronic charge; the somewhat younger sub-field known as spintronics that it spawned focuses, as you might guess from the name, on the exploitation of spin. Spintronics and nanotechnology are inherently interrelated. We've already seen that device miniaturization has pushed silicon technology to the nanoscale. Similarly, spintronics devices require control and processing at the nanometre scale. A major commercial success story in spintronics throughout the first decade of this century—namely, the hard drive at the core of the Apple iPod and other MP3 players—relied on 'sandwiches' of materials whose layers were just nanometres thick.

The dramatic increase in storage capacity that fuelled the rapid rise of MP3 players—'1,000 songs in your pocket', as Apple put it at the time—was due fundamentally to the exploitation of a spin-related effect known as giant magnetoresistance (GMR). Discovered independently, but in parallel, in the 1980s by research groups led by Albert Fert of Université Paris-Sud, Orsay, and Peter Grünberg of the Jülich Research Centre, GMR involves harnessing spin to control the flow of electrons through a device. Spins both generate a magnetic field and control the resistance to electrical current; a 'spin-up' electron can more easily pass through a material with the same electronic spin orientation as compared to its 'spin-down' counterpart. This means that tiny changes in magnetism can give rise to very large changes in electrical resistance, hence 'giant magnetoresistance'. Fert and Grünberg fabricated samples that comprised stacks of ultrathin (~ nanometres thick) layers of alternating ferromagnetic and what are known as antiferromagnetic materials (iron and chromium, to be specific). You will be very familiar with ferromagnetic materials—any fridge magnet you've encountered is a ferromagnet: the electron spins line up so that there's a net magnetization. In an antiferromagnet, however, the spins pair up anti-parallel. In other words, every spin-up electron has a spin-down partner. This means that while the material still exhibits magnetic properties—because electron spin plays a central role—there is no net magnetization.

Both of the research teams led by Fert and Grünberg, respectively, were surprised to see much larger changes than expected in the electrical resistance of the samples in response to relatively weak magnetic fields: they had discovered the GMR effect, for which they won the Nobel Prize in Physics in 2007. In a relatively rare example of what is known as the linear model of innovation in action—where fundamental, curiosity-driven, 'blue skies' research such as that pursued by Fert and Grünberg stimulates technological innovation—IBM Almaden scientist Stuart Parkin quickly appreciated the potential of GMR for increasing the capacity of hard disks. (The translation of fundamental discoveries to commercial product is almost always not this linear and generally involves a complicated feedback loop of basic and applied research, engineering, and economic considerations.) In 1991, Parkin and colleagues filed a patent for what they called a 'spin valve', a device based entirely on GMR. It was this technology that underpinned the rise of not just the iPod but data storage capacities measured in terms of (the now mundane) GB rather than MB. And at the core of it all is nanotechnology.

The single spin limit

GMR devices of the type designed and fabricated by Fert, Grünberg, Parkin, and colleagues (and many other groups since their pioneering work) involved nanoscale magnetic layers, which, although ultrathin, nonetheless comprised a vast, uncountable number of spins because the area of the devices was relatively large. As we've seen repeatedly throughout this *Very Short Introduction*, nanoscience and nanotechnology are frequently concerned with the control of matter at the most fundamental limits. So how far down can we push the control of spin?

Given that scanning probe microscopes can resolve atoms, can they resolve single spins? Is it possible to probe the spin state of a particular atom? Can we carry out logic operations, and fabricate logic gates, on the basis of the control of single spins? Remarkably, the answer to each of these questions is a resounding yes. As long ago as 1990, and in a highly influential study, Roland Wiesendanger and his team at the University of Hamburg demonstrated spin resolution with the STM. They exploited what is known as spin-polarized tunnelling where, just as for the GMR effect, the flow of electrons depends on the alignment (or misalignment) of their spins. In spin-polarized tunnelling, a magnetic tip is used to image a magnetic sample. Electrons in the sample whose spin state matches that of the tip (i.e. 'up' or 'down') have a higher probability of tunnelling, and thus a higher tunnel current will be measured. Just two years following their first experiments, Wiesendanger and colleagues had refined their measurements to the point where they achieved spin resolution at the atomic limit, detecting the difference in the spin state between different Fe ions in an iron oxide sample.

The Hamburg team have continued to drive inspiring advances in spin resolution and control over the decades since these pioneering experiments. But they're not alone in exploring the limits of spin control at the nanoscale and below. Once again, IBM is a key player. Spin valve technology stemmed from IBM Almaden, due to the efforts of Parkin and his team, so it is unsurprising that the Almaden Lab, and its alumni, have been a major influence on the field. This is especially true when it comes to work at the single atom and single spin limits, for which Andreas Heinrich's research group at the Centre for Quantum Nanoscience in Seoul and their collaborators (including, in particular, Chris Lutz at IBM

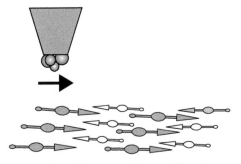

20. Artist's impression of how a scanning tunnelling microscope images spin orientation at a surface. The tip has been functionalized with a magnetic atom, molecule, or cluster so that it has a well-defined spin state that in turn interacts with the spins of the surface atoms.

Almaden) have dramatically pushed forward the boundaries. They use a technique called electron spin resonance (ESR), which in its traditional mode of operation not only involves measurements of systems comprising order 10^{10} spins but also lacks any type of spatial resolution. Heinrich and his team have radically redefined ESR by combining it with STM, enabling spin measurements with atomic resolution. As we've seen throughout this book, however, STM is both an imaging and a manipulation tool. (Heinrich's team (while he was at IBM Almaden) was responsible for the 'Boy and His Atom' stop-motion video discussed in Chapter 2.) Spins can therefore be moved, with atomic or (sub)molecular precision, to investigate how they interact. Moreover, the tip can be used to adjust the local magnetic field applied to a spin system (be that an atom or a molecule) (Figure 20).

Quantum computing and qu-bits

Spin forms the basis of many models of quantum computing, and, in a ground-breaking experiment in 2019, Lutz, Heinrich, and their teams demonstrated the fabrication, imaging, and manipulation of single-atom qu-bits. The essential idea is that while, classically, a binary bit—that is a simple two-state system—can be in either the '0' or '1' state, in quantum mechanics we can have a superposition, or, in plainer language, simply a mixture of those two states. We label the two states (using a notation due to the physicist Paul Dirac) as $|0\rangle$ and $|1\rangle$ and we can, in principle, have any mixture of those two states we like, as long as the sum of the probabilities for measuring the states always equals 1. Mathematically, our overall quantum state, which we traditionally denote using the Greek letter ψ, can be written as follows:

$$|\psi\rangle = c_1|0\rangle + c_2|1\rangle$$

where c_1 and c_2 are coefficients that tell us how much of the base $|0\rangle$ and $|1\rangle$ states contribute to the overall quantum state.

Superposition is too often painted as yet another weird quantum effect with no analogue in the real world. This is frustrating because, for one, every time a guitar string is plucked, it is the superposition of the various modes of vibration of the string, that is its harmonics, that defines the overall sound. This phenomenon is, of course, not just limited to guitars, or, indeed, other musical instruments—it is the bedrock of a great deal of science and engineering in the macroscopic, classical world. At the nano and quantum scales, superposition similarly involves the summation—that is the mixing—of different waves.

Lutz's team, led by Kai Yang, used the spin of single titanium atoms adsorbed on a very carefully chosen surface as the physical realization of a qu-bit. By applying high frequency radio waves—microwaves—from an STM tip, they could control the overall spin direction and 'dial in' the particular superposition state they required. These single atom qu-bits are also extremely sensitive to their environment, including the presence of neighbouring titanium atoms. Yang and colleagues therefore used the STM to precisely position atoms, enabling them to establish entangled qu-bits: a quantum system where the state of one qu-bit is so entirely dependent on the other that they are inextricably entwined.

This is all very exciting and revolutionary science but, again, there is the difficult issue of the choice of substrate and associated processing. In order to decouple the qu-bits from their environment to as large an extent as possible, the titanium atoms are adsorbed on a highly specialized substrate: a thin layer of magnesium oxide on a silver sample. That platform is not well suited to a scalable technology (but the motivation for Yang et al.'s elegant experiments was not a technological development of this type in any case).

Michelle Simmons's team at the University of New South Wales, and their collaborators, have instead focused on realizing qu-bits, and the associated quantum computing architecture, in silicon. In a *tour-de force* implementation of a series of innovations in nanotechnology over the last 20 years or so, Simmons, colleagues, and collaborators have taken the hydrogen-removal technique to its very limits. (See Figure 9 for just one example.) They use STM tip-induced desorption of a single H atom to produce a reactive site on an otherwise hydrogen-terminated silicon surface, which is subsequently exposed to a gas of phosphorus-containing molecules. Those molecules attach at the reactive site, and with the appropriate control of the exposure parameters it's possible to introduce a single phosphorus atom at a specific, pre-defined atomic site defined by the removal of hydrogen. Silicon is then deposited over the top to 'bury' the phosphorus atom, so that it is incorporated in the crystal lattice.

These innovations, and the overarching UNSW quantum computer project, have as their core motivation the fabrication of a QC architecture known as the Kane model. Briefly, the Kane model involves the control of the coupling of nuclear and electron spin for the phosphorus atoms embedded in the matrix. These form the qu-bits and, just as with the titanium atoms exploited by the IBM Almaden team, careful control of the separation of individual phosphorus atoms in the silicon matrix will enable the degree of coupling and entanglement to be fine-tuned.

A silicon-based solid state quantum computer is an exceptionally challenging target. Nonetheless, a spin-off company from the UNSW initiative, Silicon Quantum Computing, has as its ultimate, long-term objective that it will enable 'access to useful quantum computing solutions for a broad audience of users and multiple uses by the mid-2030s'. If Silicon Quantum Computing can maintain this timeline, commercial silicon-based quantum computing will be with us much sooner than even many of its key proponents ever imagined.

This chapter has been heavy on top-down, precisely engineered, and largely inorganic approaches to nanotechnology. In the next, we redress that balance and see just how nature harnesses randomness at the nanoscale to power molecular machines.

Chapter 5
Nanomachines

Late afternoon, 28 April 2017. A joint Austrian–US team comprising nanoscientists from the University of Graz and Rice University, Houston has just won the world's first NanoCar Race, a nanoscale Grand Prix involving a total of six international teams. Their nanoscopic car (see Figure 21) achieved a record-breaking speed of 100 nm per hour on average, but sometimes reached the dizzying heights of 300 nm/hr. (Although this is impressive from the point of view of top-down control of molecular trajectories, it's perhaps worth noting for context that if a Formula 1 car travelled at this speed it would take roughly 600,000 years to complete a single lap of a Grand Prix track.)

Each competing research group had designed and synthesized its own single molecule nanocar, which had subsequently been driven around a racetrack—a metal surface—using the tip of a scanning tunnelling microscope. Unlike the examples of atomic and molecular manipulation we've seen in previous chapters, however, pushing the molecule with the apex of the STM tip was strictly prohibited by the rules of the race. Instead, the molecule had to be propelled without direct mechanical—or, to be more technically correct, chemomechanical—contact.

There were two primary sources of 'fuel' for the nanocar, and both could in principle act in tandem. First, the injection of electrons

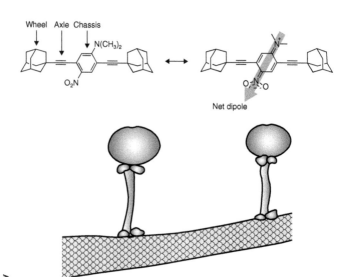

21. Top: Chemical structure of Dipole Racer, the winner of the first NanoCar race. The wheels are adamantane ($C_{10}H_{16}$) molecules. Bottom: Artist's impression of kinesin, a motor protein.

from the STM tip could be used to excite molecular diffusion via a process known as inelastic tunnelling. This involves the same quantum mechanical tunnelling effect that is at the core of the STM's operation but instead of tunnelling without losing energy, as accounts for the vast majority of the tunnel current, the electrons instead excite molecular vibrations and therefore lose energy; this is why the process is described as inelastic. Alternatively, or in parallel, the car could be driven by the influence of the electric field from the STM tip. Assuming that the molecule has a large dipole moment, or its electronic charge is highly polarizable, that is it can be readily distorted by an electric field, then it will respond to the high electric field due to the tip.

The Austria–US team raced home in 1 hr 33 mins, far outstripping the competition, who, in many cases, dropped out long before reaching the finish line. Their success was due both to very clever nanocar design principles and associated synthesis strategies which took account of the surface-molecule interactions that needed to be exploited and surmounted, as well as a new approach to STM-driven molecular manipulation that relied only on measurements of tunnel current, circumventing the highly time-consuming acquisition of images between manipulation events.

Impressive though the winning nanocar may be, it's still the case that it is driven in an entirely top-down fashion: humans are guiding it almost every step of the way. Left to its own devices, the nanocar would simply diffuse randomly around the surface, with no bias for a particular direction. Yet the natural world is awash with molecular machines that achieve unidirectional motion with no need for an external guiding intelligence; as you read this, every cell of your body is teeming with nanoscale molecular machinery. Nature has harnessed nanotechnology almost from the start of life on Earth to perform a variety of essential functions, and the sophistication, intricacy, and elegance of its nanomachinery far outstrips anything that scientists are (currently) capable of creating.

Nature's nanomachines

Pictured alongside the STM-powered nanocar shown in Figure 21 is a biomolecular motor, from the kinesin family, that leaves its synthetic counterpart in the dust in very many ways. Completely autonomous, unidirectional, extremely fast (routinely achieving 2,000 nm per *second*), and remarkably energy efficient, kinesins, and other similar motor proteins, are central to a wide array of essential processes in cells including movement, division, and transport of sub-cellular structures such as vesicles, organelles, and neurotransmitters. (Note that there isn't just one form of kinesin—there are at least 45 different types in humans alone.) Those processes in turn underpin and drive some rather fundamental aspects of our being. It's no exaggeration to say that without motor proteins we'd be immobile: motor proteins are at the core of muscle activity. Indeed, we wouldn't be here in the first place without motor proteins; in the absence of kinesin and its counterparts our cells wouldn't develop and we'd fail to reach even an embryonic stage of development. And when motor proteins fail in humans, the effects on health are severe: cancer, neurodegenerative disorders, and polycystic kidney disease have each been linked to deficiencies in the behaviour of biomolecular motors. Motor proteins are thus very much the engines of life.

Kinesin, whose motor unit is approximately 8 nm across, converts chemical energy into mechanical movement (like all other motor proteins), transporting its molecular cargo along tracks known as microtubules. As such, kinesin is perhaps best described as a nanoscale train rather than a nanocar, but there the comparison ends. Astoundingly, kinesin walks, rather than rolls, along the tracks of the microtubule. While the protein was discovered as recently as 1985 by Ron Vale (of the Marine Biological Lab, Massachusetts at the time), Thomas Reese (University of Connecticut), and Michael Sheetz (Stanford), it took almost another 20 years to decipher just how kinesin walks along its track. In fact, it's more accurate to describe kinesin's motion as limping—an exceptionally speedy limp (as compared to the nanocar) of 2,000 nm/s, but a limping motion nonetheless.

Although the precise details vary across the kinesin family, the overall structure comprises two heavy chain molecules that form a pair (a molecular dimer), which in turn link to two light chain molecules that are specific for different cargoes. Each heavy chain consists of a globular protein head—the motor unit—which is connected to a 'stalk' that terminates in a carboxy group, enabling the connection of a light chain. This short description doesn't begin to do justice to the complexity of the structure—it's a remarkably impressive example of natural, bottom-up, nanoscale bioengineering. What's of most interest in the context of this *Very Short Introduction* is the motor unit.

Powered by adenosine triphosphate (ATP), the conformation of the motor unit, that is the configuration of its chemical bonds, is changed by the binding of ATP and its subsequent hydrolysis to adenosine diphosphate (ADP). (Conversion of ATP to ADP in this way is not unique to kinesin and other motor proteins—it's the mechanism of energy conversion in all living cells.) It is the conformational change of the motor unit that ultimately drives the motion of the entire kinesin complex and its cargo. Unlike the nanocar, however, there is no intervention by an external force to drive the kinesin in one direction. Why, and how, does it move in just one direction?

Random, stochastic motion is everywhere in nature. Air molecules in the room around you have their trajectories 'scrambled' billions of times each second due to intermolecular collisions. Although it's technically possible for all of those collisions to produce a correlated motion of the gas molecules so that they all travel in the same direction, it is overwhelmingly improbable—by a factor that is much greater than the total number of atoms in the observable universe—for this to happen. Echoing the description of perfume diffusion in the previous chapter, the equilibrium state of the system is that the air molecules follow random trajectories. Similarly, in the liquid phase (including, in particular, the *in vivo* environment of a living cell), molecules will follow random trajectories giving rise, for one, to Brownian motion of larger particles/organisms constantly buffeted by the surrounding medium. The physics and chemistry of the nanoscale means that it's a sticky, gooey, and bumpy environment inside a cell: sticky because of the ubiquity of van der Waals and dispersion forces giving rise to intermolecular attraction; gooey because viscosity—a measure of the resistance of a fluid to flow—dominates at nanometre length scales; and bumpy because of the continual pummelling of the surrounding molecules. In a memorable description of life at the nanoscale, the physicist R. Dean Astumian has noted that molecular machines must 'swim in molasses and walk in a hurricane'.

How, then, does kinesin—or any other molecular motor—walk in one direction? Why don't molecular motors follow a random trajectory, like other molecules, due to the tumultuous environment in which they have to work? The precise biophysics underpinning kinesin's ability to move in one direction has yet to be fully elucidated, but one school of thought focuses on an important model of directed motion in a wide range of biomolecular processes: the Brownian ratchet. This is a mechanism for rectifying random (Brownian) motion so as to produce a net displacement or force in a given direction. Short-range intermolecular attraction is exploited to trap a system—be it a motor protein, an enzyme, or any one of a wide variety of other biomolecular components—after a random fluctuation pushes it in the right direction. George Oster, an American mathematical biologist, described Brownian ratchets as Darwin's motors for the following compelling reason:

> In a broader sense, the idea of generating order by 'selecting' from random variations is hardly new—it is the fundamental idea of Darwin's theory of natural selection. In the context of motor proteins, the 'order' created is a directional force, and the agents of selection are intermolecular attractions.

A schematic illustration of the operation of a ratchet potential is shown in Figure 22. A collection of molecules is originally trapped in a potential well, whose depth and shape are controlled by the structure and energy of the biochemical environment. When the potential is switched off, the molecules are free to diffuse away from their original site. If, however, the potential is switched back on in a time scale that is sufficiently short so that the molecules don't have time to diffuse appreciably, they will follow the gradient of the potential 'downhill' and some will get trapped in the well to

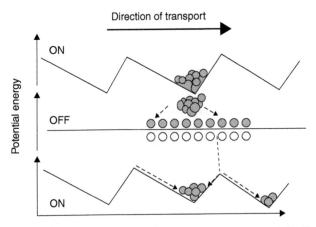

22. The basic operating principle of a Brownian ratchet. By periodically switching the potential on and off, it is possible to produce directional motion. The molecules move in the direction of decreasing potential, just as a ball rolls down a hill.

the right. If this process is repeated over and over the net result is a motion of the molecules to the right.

When it comes to kinesin, an individual molecular machine, the principle is broadly the same. Each motor unit of kinesin has two binding sites—one for the microtubule and another for ATP. Moreover, in a single step one of the heads remains connected to the microtubule, while the other moves; on the next step the roles are reversed such that the motor moves forward via a hand-over-hand mechanism. (It's a little like climbing a ladder with just your hands—both hands can't release together or you'll fall down.) As noted above, the binding of ATP triggers a conformational change; this in turn releases the trailing motor unit, which in turn swings forward and uses thermal energy (Brownian motion) to explore space and find the lowest energy binding configuration to the microtubule that it can. Then the cycle begins afresh, and the trailing motor unit swings forward for the next step.

It's an exceptionally elegant strategy—randomness and fluctuations are harnessed to produce directed motion. Given that evolution has fine-tuned molecular motors over the millennia until they are exceptionally efficient at converting biochemical to mechanical energy, nanotechnologists have realized that starting from scratch in the design of nanomachines to carry molecular cargoes would be too much like reinventing the wheel. So extensive effort has been invested in stealing from nature to embed molecular machinery in artificial nanosystems.

Biomimetic and biokleptic nanotechnology

To the very best of my knowledge, the term biokleptic was coined by the physicist (and nanoscientist) Richard Jones, of the University of Manchester. Jones draws a distinction between biomimetic and biokleptic nanotechnology; the first is inspired by nature's design principles at the nanoscale, whereas the latter unashamedly steals wholesale from the natural world—molecular motors (or components thereof) are imported in entirety into a synthetic or artificial nanosystem. When it comes to molecular machinery, nanotechnologists are both mimic and thief.

Biomolecular motors such as kinesin, dynein (a motor protein that walks along microtubules in the opposite direction to kinesin), and myosin (which drives muscle contraction) have been integrated with a variety of artificial actuators and sensors. Merging natural and artificial biotechnology is immensely challenging, although impressive progress has been made in the fabrication of, for example, artificial cilia. Composed of an array of microtubules, cilia play a vital role in human physiology and are found in the lungs, respiratory tract, middle ear, kidney, eye, and sperm (whose flagellum is a modified cilium). Motile, that is moving, cilia rhythmically wave (or 'beat'), and in the context of our respiratory system are responsible for keeping airways clear of mucus and dirt. The exploitation of the biological principles underpinning the behaviour of cilia is important not only from a biomedical perspective, however. Biomimicry of this type also enables the design of what is now known as active matter: a system or material comprising a large number of active agents, operating out of thermal equilibrium so as to exert mechanical forces.

In 2018, a team of researchers at Hokkaido University and the Tokyo Institute of Technology led by Akira Kakugo fabricated artificial cilia by attaching microtubule-kinesin units to polystyrene beads, all contained within a flow cell through which they could stream different chemicals. When the kinesin was exposed to a solution of ATP, the artificial cilia showed a beating motion whose frequency was tunable by changing any one of a number of experimental parameters, including the density of kinesin along the tubules and their length. Although bionanotechnology has not yet advanced to the point where this type of architecture could be exploited in biomedical applications such as prosthetic limbs, the groundwork is certainly being laid.

Bionano machinery can also be used for sensing applications. Although there is a strong focus in state-of-the-art biosensing in what's known as microfluidic and nanofluidic technology—where very small volumes of the substance(s) to be analysed are injected into arrays of nanoscopic capillaries or pores—an alternative approach is instead to introduce so-called smart nanosensors into the sample. A collection of these sensors is known as smart dust and is as close as nanotechnology has got to that staple of nanotechnology science fiction: the nanobot.

Smart, self-powered, and effectively sentient, the nanobot of science fiction tends to be envisaged as the *Nautilus* (from Jules Verne's *20,000 Leagues Under the Seas*) writ small: a nanoscopic submarine that propels itself through our bloodstream, zapping anything untoward that it finds. Nature, however, hasn't produced anything that looks quite like a scaled-down submarine, that is a miniaturization of macroscopic technology, because the physical, chemical, and engineering principles required for efficient motion at the nano (and/or micro) scale are very different from those at play in our everyday world. Nanomachines must indeed swim in molasses and walk in a hurricane, and this environment means that simply down-scaling traditional engineering principles will not be efficient. Nonetheless, it is certainly possible to extract and exploit biomolecular devices in artificial, inorganic nanosystems so as to develop scaled-down versions of macroscopic technology. As long ago as 2000, a team of Cornell researchers fabricated a hybrid nanomechanical device powered by a biomolecular motor (an enzyme known as F_1-ATPase) that drives a nanoscale propeller.

While nanobots of the type envisaged in science fiction—including *Dr Who*'s nanogenes, *Star Trek*'s nanites, the terrifying swarm of Michael Crichton's *Prey*, and the so-called grey goo inspired by Eric K. Drexler's writings—will remain fictional for quite some time to come, autonomous nanoscale agents are nevertheless being developed by a number of research groups. In 2019, Thorsten Fischer, Ashutosh Agarwal, and Henry Hess of the University of Florida developed a smart dust biosensor that exploited kinesin in order to shuttle microtubules exposed to a target analyte between different locations. This enabled tagging (with fluorescent molecules that act as markers) and detection to be carried out as spatially separated processes. Although very much in its infancy, this type of technology integrates microscale and nanoscale biomachinery with artificial, inorganic engineering and is very likely to play a defining role as nanotechnology matures in the 21st century.

Engineering artificial nanomachines: chemical topology

Thus far, the examples of biomachinery I've chosen have been skewed towards the biokleptic side of the engineering space. In parallel, however, there has been impressive progress in the realization and implementation of artificial nanomachines, both externally powered (as in the nanocar) and autonomously driven. In particular, the Nobel Prize in Chemistry was awarded in 2016 to Jean-Pierre Sauvage, Fraser Stoddard, and Bernard (Ben) L. Feringa for their pioneering work on the design, synthesis, and development of molecular machines. As the Royal Swedish Academy of Sciences describe in the scientific background to the 2016 prize in chemistry, two major, and very much interlocking, advances have underpinned the realization and development of artificial molecular machinery: chemical topology and the exploitation of isomerizable bonds.

Topology—broadly, the study of objects that are stretched, twisted, crumpled, and/or knotted—plays an exceptionally important role in biochemistry and served as a core inspiration in the work of Sauvage, Stoddard, Feringa, and their respective research teams. Knots are prevalent throughout biological systems. Figure 23 shows an example of a knotted DNA structure that has been created by what are known as topoisomerases, enzymes that have been described as the magicians of the biomolecular world because they allow DNA strands and double helices to, in effect, pass through each other via breaking and remaking bonds, or to tie/untie knots along the chains. (Malfunctioning topoisomerases are thought to contribute to a variety of forms of cancer.)

Sauvauge and Stoddard realized that chemical topology could be exploited in the generation of molecular systems with interlocking components, whose relative positions are controlled via photochemical, electrochemical, and/or mechanochemical stimuli. These systems would have all the ingredients required for nanoscale machines, mimicking the action of not only naturally occurring nanomotors such as kinesin but also a much broader range of biomolecular machinery. In 1983, Sauvage and his co-workers at CNRS, Louis Pasteur University, Strasbourg, established a new technique that enabled a much more straightforward synthesis of two key molecular classes involving interlocking, but movable, units—the catenanes and rotaxanes.

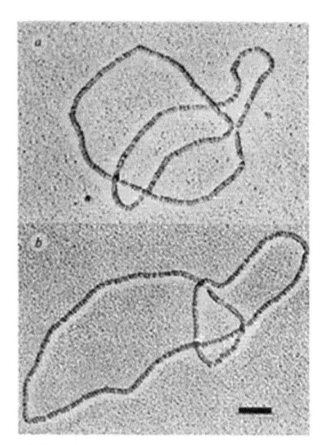

23. A DNA knot.

As can be seen in Figure 24, the catenanes comprise two interlocked rings, whereas the rotaxanes are, in essence, a molecular wheel-and-axle unit where the wheel is trapped on the axle by two large stoppers at each end. In addition to its ability to rotate, however, the wheel—or to use the correct chemical term, the macrocycle—can move back and forth along the axle.

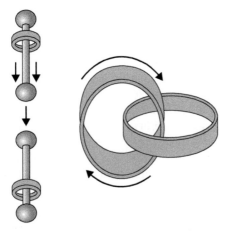

24. Schematic illustrations of the molecular classes known as (left) rotaxanes and (right) catenanes.

Sauvage and colleagues' new technique allowed not just for the synthesis of catenanes and rotaxanes, but a wide array of topologically complex molecular systems involving knots, links, and locks. Throughout the nineties, Stoddard, Sauvage, and co-workers carried out a series of pioneering advances in the control of both the translation and rotation of molecular sub-units in catenane and rotaxane systems, spanning photochemical, electrochemical, and thermal actuation. At the turn of the millennium, Sauvage's team synthesized a rotaxane structure capable of expansion and contraction—a synthetic analogue of the biomolecular motors underpinning muscle motion. This was closely followed (in 2004) by Stoddard et al.'s synthesis of a rotaxane structure that could bend a nanoscopic gold cantilever using a similar 'molecular muscle' approach.

The third recipient of the 2016 Nobel Prize in Chemistry, Ben Feringa, pursued a rather different, but no less pioneering and innovative, strategy in the development of molecular machinery. In 1999, Feringa's team at the University of Groningen in the Netherlands, in collaboration with researchers at Tohoku University, fabricated the first synthetic molecular motor capable of unidirectional motion using an ingenious and elegant approach known as bond isomerization. They achieved unidirectional motion by essentially reconfiguring the chemical bonds of a molecule, a step at a time, in a manner that ensured that the motor could only rotate one way. In the intervening years since the synthesis of that first motor, Feringa, colleagues, and collaborators have improved the design dramatically, to the point where a rotation frequency of 12 MHz was demonstrated in 2014.

Progress in molecular machinery has not been limited to Nobel Prize winners and their teams. David Leigh's research group at the University of Manchester (and, previously, Edinburgh), for one, not only has played a central role in the development of novel strategies for molecular machines but also, as with the work of Winfree, Rothermund, Seeman, et al. on DNA nanotechnology, has explored the deep links between information, energy, entropy, and patterns at the nanoscale. Once again, molecular topology— knots, chains, and links—has been an enduring inspiration for the Leigh group's work. As Leigh himself put it in an interview with *Chemistry & Engineering News* in 2020,

> 'Knotting and weaving have always had great technological impact for humans,' Leigh says, noting that the invention of knots and weaves helped humans create weapons, tools, nets, and cloth. 'Who's to say it won't be the same for molecular structures?'

Leigh's team (along with collaborators across the University of Manchester) holds the Guinness World Record for the finest fabric ever woven. They chemically wove molecular threads in two dimensional layers to form a nanostructured fibre whose thread count—that is the number of strands per inch—was of the order of

50 million, as compared to the thread count of ~ 1,500 for fine Egyptian linen. For those who prefer SI units, this equates to threads that are 4 nm across.

In a series of advances over the last 20 years or so, Leigh and colleagues have developed an impressive array of molecular machines including nanomotors, walkers, information ratchets—light-powered molecular machines that transfer information rather than matter—and, most recently, a programmable molecular robot. A nanobot, in other words. So many dystopian visions of our nano-enabled future—not least Michael Crichton's *Prey*, to which I referred earlier—focus on a scenario whereby programmable, and ultimately sentient, nanobots run amok. Should we be scared?

Chapter 6
Are the nanobots nigh?

'Tea. Earl Grey. Hot.' And, as if by magic, Captain Picard's beverage of choice appears out of thin air from a replicator in his room on the *Starship Enterprise*. The author Arthur C. Clarke asserted that any sufficiently advanced technology is indistinguishable from magic. *Star Trek*'s replicator technology certainly seems magical to 21st-century eyes: a compact device that can conjure up virtually any material or object—including food, drink, air to breathe, clothing, and medicine—by assembling it from its raw components. In *Star Trek*, that raw material is pure energy and it's not at all clear just how the energy–matter conversion process takes place in a 'desk top' unit; it's science fiction, after all.

But what if we *could* deconstruct matter into its component atoms or molecules, and then build it back up again into a completely different form, and to a pre-defined blueprint? That sounds hopelessly optimistic when considered in the context of replicator technology. And yet chemists do this all the time. The variety of synthetic nanostructures we've encountered throughout this *Very Short Introduction*, from NanoKid to molecular motors, are, after all, a result of just that type of matter manipulation: raw chemical ingredients are broken down via chemical reactions and their constituent atoms rearranged to form a new structure. That's nothing more than traditional chemistry and we've been doing it since, arguably, the discovery of fire in our prehistory.

We therefore already have the chemical technology to rearrange the atomic and molecular building blocks of matter into other precisely defined forms. So what is it about the replicator that makes it appear so 'magical' to our 21st-century perspective? Well, for one, it's conjuring matter out of thin air—or whatever nebulous form of 'energy' is being invoked in the *Star Trek* universe. But it's also the 'on demand' aspect of the technology that we find so other-worldly. We're used to asking Siri a question and getting a virtually instantaneous response—sometimes it's the wrong answer, admittedly, but at least she always tries. What the replicator gives us is not information, as provided by Siri, but *matter* on demand, and configured in just the way we want.

Replicators have been compared by some to another type of futuristic technology known as the universal assembler or molecular assembler. This concept in turn stems from the widely critiqued vision of nanotechnology first put forward by K. Eric Drexler in the 1980s and described at length in his books *Engines of Creation: The Coming Era of Nanotechnology, Nanosystems*, and *Radical Abundance*. Drexler envisages a nanotech-enabled future utopia where atomically precise manufacturing (APM) carried out by molecular assemblers 'will be able to make virtually anything from common materials without labour, replacing smoking factories with systems as clean as forests'.

What a wonderful world that would be! Drexler's version of nanotech is an inspiring and encouraging vision of what might be possible if we could manufacture products with atomic precision. And what's remarkable is that the kernel of Drexler's vision, namely computer-controlled chemistry with not just single atom but *single chemical bond* precision, is now carried out in a considerable number of nanoscience laboratories across the world (as we've seen throughout previous chapters). I hesitate to call bond-by-bond manipulation of matter routine just yet but it's certainly a well-established technique for many SPM research groups.

Intense debates raged across the nanoscience community throughout the 1990s and 2000s concerning the likelihood of universal assemblers becoming a reality and the ultimate capabilities, potential, and dangers of nanotechnology's ability to assemble matter, atomic bit by atomic bit. Echoes of those debates are still clearly heard today, not least with regard to the human–machine interfaces being developed by Elon Musk's Neuralink company.

Despite his prescience in predicting that chemical reactions could be carried out with atomic precision under computer control—and this was a number of years before the invention of the first scanning probe microscope in the early 1980s—Eric Drexler was widely criticized and castigated by not just nanoscientists but the broader scientific community. To understand why Drexler attracted the opprobrium he did, we first need to consider his universal assembler idea and associated mechanosynthesis concept. As with the *Star Trek* replicator, these remain firmly in the realms of science fiction but the operating principles are, at least, somewhat better defined: instead of a nebulous energy–matter conversion, Drexlarian nanotech involves the conversion of one form of matter into another.

This might sound a lot like 21st-century alchemy but the matter manipulation occurs at the level of chemical bonds, not atomic nuclei. We are not transmuting one element into another; there's no conversion of 'base' materials into gold. Instead, a future universal assembler would break down matter and build it up again into the new arrangement we require by chemomechanical manipulation of atoms and molecules: chemical bonds would be formed by literally forcing atoms together in the right places to form the desired product, be that a hot cup of Earl Grey, a new pair of socks, or a solid state quantum computer. Drexler variously called this process mechanosynthesis, machine phase chemistry, or molecular manufacturing. If this type of technology could be realized, it would represent the ultimate form of nanotechnology.

Drexler's machine phase nanotech is, in essence, real world engineering writ small. *Very* small. Raw materials—simple molecules such as, for example, ethyne (C_2H_2) and methane (CH_4)—are fed into a manufacturing plant, that, unlike its cavernous industrial revolution counterpart, occupies no more space in the kitchen than a modern microwave oven. Long before 3D printing was developed, Drexler envisaged a matter manipulation technology that is, in essence, 3D printing with atoms. In this type of nanofactory, molecules are bonded together to form nanoparticles, which in turn are connected to form microparticles, and they in turn are linked to create structure on sub-millimetre, and subsequently much larger, length scales; a hierarchy of Lego blocks, in other words, ultimately building up macroscopic, everyday structures like computers, cars, and houses.

Figure 25 shows two frames from an animation that was put together by a company called NanoRex in 2005 to illustrate the architecture of a nanofactory and to explain how mechanosynthesis would work. Small, simple molecules (in this case C_2H_2) are fed into the factory, broken down, and passed along a series of atomically precise wheels, cranes, and conveyor belts. Molecules and atoms are transferred and bonded, building up an ever more complex structure until, at the end of the process, out pops a laptop. No ordinary laptop, of course—this is a device that has been fabricated with single atom precision and is thus free of the deficiencies associated with the top-down approach to materials processing and semiconductor device fabrication that we covered in Chapter 3.

True, like the *Star Trek* replicator, the nanofactory depicted above is science fiction. But the key question is this: what's wrong with the molecular manufacturing concept? Does it break fundamental laws of physics or chemistry? Is it in violation of the second law of thermodynamics? Is energy conservation violated? Given that chemistry already allows us to break down matter and mix up chemical ingredients to form new compounds, why is it that

25. Two frames from an animation of the molecular nanofactory/
assembler concept put forward by K. Eric Drexler.

Drexler's extrapolation of this capability to a computer-driven
'machine phase' caused such a furore? Drexler was denounced as a
crank and delusional; his molecular manufacturing concept
dismissed as fundamentally impossible. No less a luminary in the

nanoscience community than the late Richard Smalley, a Nobel laureate (for his role in the discovery of buckminsterfullerene) and formidable (nano)chemist, dismissed Drexler's ideas as entirely unworkable and hopelessly naïve:

> Much like you can't make a boy and a girl fall in love with each other simply by pushing them together, you cannot make precise chemistry occur as desired between two molecular objects with simple mechanical motion along a few degrees of freedom in the assembler-fixed frame of reference. Chemistry, like love, is more subtle than that.

Worse, Smalley painted Drexler as a nano bogeyman, whose ideas about universal assemblers and self-replicating nanobots were not just misplaced but alarming and monstrous.

Although Drexler didn't help his case by making wildly optimistic predictions about the timeline for the development of the molecular manufacturing capability he envisaged—suggesting that it was just 'one to three decades off' in 2001—much of the criticism targeted by Smalley (and many others) was unfair and misrepresented the arguments underpinning mechanosynthesis and molecular manufacturing.

This is not to say that the Drexlerian vision of nanotech, as described in *Nanosystems*, is not open to sustained criticism. Far from it. There are many unresolved issues with the automated assembly hierarchy—from the (sub)nanoscopic to the macroscopic—that are perhaps best summed up by the physicist Wolfgang Pauli's pithy assertion: 'God made solids, but surfaces are the work of the devil.' The devilish physics and chemistry of surfaces plays a central role in *all* forms of nanotechnology and nanoscience, not just Drexler's futuristic molecular manufacturing technology; for one, as we shrink from chunks of matter that we can hold in our hands to the tiniest of nanoclusters comprising a countable number of atoms, the surface-to-volume ratio increases dramatically. In many ways, surface science and nanoscience are effectively synonymous; indeed, with the rise of nanotechnology in the 1990s, many surface scientists rebranded themselves as nanoscientists.

Notwithstanding these difficulties with Drexler's molecular manufacturing framework, and despite (or perhaps because of) the surrounding hype, his vision has clearly had a substantial influence on at least some nanoscientists (including, it has to be said, myself.) While the Leigh Group's molecular robot (or nanobot) technology, briefly mentioned in the previous chapter, is vastly different from that envisaged by Drexler, they reference his ideas in describing the background and motivation to their work. It appears that he was at least in part responsible for inspiring their (and others') research. Nonetheless, their research backs up Philip Ball's assertion that 'Drexler ends up proposing to do the hard way things that might be more readily (and more quickly) accomplished with a little chemical ingenuity'. We're still a long way from having to worry about self-replicating nanobots reducing us all to what was memorably described by Prince Charles as grey goo. Such remarks, coupled with media scare stories, contributed to an unwarranted public distrust of nanotechnology that cast a long shadow. There are much more pressing technological and sociopolitical challenges that society has to face—climate change foremost among them (a problem to whose solutions nanotechnology certainly contributes)—than an infestation of nanobots.

Nonetheless, it would be remiss of me to underplay the realistic concerns expressed by many nanoscientists about the adverse effects on health and potential ecological damage that can be wrought by nanostructured materials. Christie Sayes, Associate Professor in the Department of Environmental Science at Baylor University, and her team have spent decades studying the effects of nanoparticles, nanotubes, and other nanostructured matter on living tissue and animal systems (including humans). Although it has frequently been hypothesized that nanomaterials may involve novel toxic effects entirely distinct from those of other, more traditional, materials, Sayes points out that the research published in the literature doesn't support this view. Instead, the enhanced toxicity of a nanomaterial arises because its smaller size means that it can access cells and tissues in a way that microparticles and larger particulates cannot:

> For instance, nanoparticles, when they are aerosolized and you breathe them in, they're able to reach the distal areas of the lung, where larger-sized particles are not able to deposit or reach. But the extent of the toxicity or the dose needed to elicit an adverse response is a lot lower when you're exposed to a nanomaterial as opposed to a bulk size or micro-sized particle.

In other words, it's a question of exposure and extent: it takes fewer nanoparticles to produce the same response as larger particulates. And once again, surfaces play a central role. In a highly cited and influential paper, Sayes and colleagues found that relatively subtle changes to the surface of buckminsterfullerene (involving the addition of chemical groups to make the molecule more water soluble) had an exceptionally large effect on its toxicity in human cell lines. Remarkably, the lethal dose of fullerene changed by *seven orders of magnitude* for untreated buckyballs as compared to their chemically altered counterparts. Moreover, the toxicity arose from disruption of cell membranes. Many other nanoparticles can similarly penetrate and/or disrupt biological membranes, moving through cells and tissues to cause biochemical damage and accelerate the progression of disease.

In this context, swarms of sentient nanobots are not a pressing concern. The nanotoxicity research community instead focuses on genuine health and environmental issues rather than devoting their time to tackling an existential threat that right now remains firmly in the realms of science fiction. Nonetheless, at the time of writing, there is a rapid growth in research that bridges nanotechnology and artificial intelligence, integrating machine learning with a broad range of problems involving the control of matter at the nanometre, molecular, and atomic scales. But such work is emphatically unlikely to lead to the type of nanotechnological dystopia envisaged by the futurist and entrepreneur Ray Kurzweil:

> Around 2030, we should be able to flood our brains with nanobots that can be turned off and on and which would function as 'experience beamers' allowing us to experience the full range of other people's sensory experiences...Nanobots will also expand human intelligence by factors of thousands or millions. By 2030, nonbiological thinking will be trillions of times more powerful than biological thinking.

I can confidently predict that by 2030 our brains will not be flooded by nanobots, beaming experiences from one person to another. Instead, researchers will be continuing to embed machine learning algorithms in rather less fanciful, much more ethical, and substantially less dystopian aspects of nanotechnology, including image and spectral classification in various forms of microscopy, automation of time-consuming tasks such as optimization of the tip of an STM or AFM, and the positioning of individual atoms and molecules.

The latter is already happening. A pioneering paper, 'Autonomous robotic nanofabrication with reinforcement learning', from Christian Wagner and colleagues at Forschungszentrum Jüelich, was published in 2020, in which the authors describe how a scanning probe microscope has been taught to pick up single molecules with no human input. At the time of writing this represents the state of the art in the integration of artificial neural networks with nanotechnology but this just scratches the surface of what's possible. The coming years are likely to see hybrid machine learning–nanotechnology approaches quickly evolve until they are no longer a niche component but *de rigueur* in the imaging, manipulation, and spectroscopy of matter at the nanoscale.

Although nanobot swarms will therefore not descend upon us any time soon, nanotechnology is nonetheless enabling the control of matter in ways we could not have envisaged even a decade ago. On a time scale of not much more than a generation, we have progressed from a mindset where pushing a single atom was considered by mainstream science as a capability that would forever remain a *Gedankenexperiment* to the controlled assembly of nanostructures a chemical bond at a time. From this perspective, the science is even more thrilling than the science fiction.

Further reading

Chapter 1: Welcome to NanoPut

Synthesis of anthropomorphic molecules: the NanoPutians,
Stephanie H. Chanteau and James M. Tour, *Journal of Organic Chemistry* **68**, 23, 8750–66 (2003)

Nanoscience and Nanotechnologies: Opportunities and Uncertainties,
The Royal Society and the Royal Academy of Engineering;
<https://royalsociety.org/-/media/Royal_Society_Content/policy/publications/2004/9693.pdf>

Scanning Probe Microscopy: From Sublime to Ubiquitous, American Physical Society collection of influential SPM papers. <https://journals.aps.org/prl/scanning-probe-microscopy>

Positioning single atoms with a scanning tunnelling microscope,
D. M. Eigler and E. K Schweizer, *Nature* **344**, 524–6 (1990)

Feynman's fancy, Philip Ball, *Chemistry World* 8 Jan. 2009.

There's plenty of room at the bottom, R. P. Feynman, *Engineering and Science* **23**, 22–36 (1960). <https://calteches.library.caltech.edu/1976/1/1960Bottom.pdf>

Van der Waals interactions and the limits of isolated atom models at interfaces, S. Kawai et al., *Nature Communications* **7**, 11559 (2016)

Perplexed by Pauli, P. Moriarty, <https://muircheartblog.wpcomstaging.com/2014/08/15/perplexed-pauli/>

Chapter 2: The quantum, confined

A Boy and His Atom, <https://www.youtube.com/watch?v=oSCX78-8-q0>

Confinement of electrons to quantum corrals on a metal surface, M. F. Crommie, C. P. Lutz, and D. M. Eigler, *Science* **262**, 218 (1993)

For much more on the deep links between the physics of music—in particular, heavy metal—and quantum mechanics, see *When the Uncertainty Principle Goes to 11*, P. Moriarty (Ben Bella Books, 2018)

Quantum rings engineered by atom manipulation, Van Dong Pham, Kiyoshi Kanisawa, and Stefan Fölsch, *Physical Review Letters* **123**, 066801 (2019)

Realization of a particle-in-a-box: electron in an atomic Pd chain, N. Nilius, T. M. Wallis, and W. Ho, *J. Phys. Chem. B* **109**, 20657 (2005)

Michael Faraday's gold colloids, <https://www.rigb.org/our-history/iconic-objects/iconic-objects-list/faraday-gold-colloids>

<https://www.lancaster.ac.uk/news/the-nano-guitar-string-that-plays-itself>

Welcome to Clusterworld, Richard Palmer, *New Scientist* 22 Feb. 1997

Chapter 3: Tearing it down, building it up

ENIAC in Action: Making and Remaking the Modern Computer, Thomas Haigh, Mark Priestley, Crispin Rope, William Aspray, Thomas J. Misa (MIT Press, 2016)

Crystal Fire: The Birth of the Information Age, Michael Riordan and Lillian Hoddeson (W. W. Norton & Company, 1997)

Atomic-scale desorption through electronic and vibrational excitation mechanisms, T.-C. Shen et al., *Science* **268**, 1590 (1995)

An Atomic Christmas Tree, Sixty Symbols— <https://www.youtube.com/watch?v=gRF9hM_eFPU>

Drying mediated self-assembly of nanoparticles, Eran Rabani et al., *Nature* **426**, 271 (2003)

Disorder—a cracked crutch for supporting entropy discussions, Frank Lambert, *Journal of Chemical Education* **79**, 187 (2002) Lambert's other engaging and influential papers on the subject of entropy are thoroughly recommended and can be sourced via his Wikipedia page.

Random tiling and topological defects in a two-dimensional molecular network, M. O. Blunt et al., *Science* **322** 1077 (2008)

Digital alchemist seeks rules of emergence, interview with Sharon Glotzer, *Quanta* 2017

<https://www.quantamagazine.org/digital-alchemist-sharon-glotzer-seeks-rules-of-emergence-20170308/>

Self-organization: the fundament of cell biology, Roland Wedlich-Söldner and Timo Betz, *Philosophical Transactions of the Royal Society B* **373**, 20170103 (2018)

The beauty and utility of DNA origami, P Wang et al., *Chem* **2**, 359 (2017)

Transcript of Konstantin Novoselov's Nobel Prize speech <https://www.nobelprize.org/prizes/physics/2010/novoselov/lecture/>

Chapter 4: It from bit, bit from it

Quotation from Ada Lovelace, *Scientific Memoirs Selected from the Transactions of Foreign Academies of Science and Learned Societies* (1843), Article XXIX

Information, physics, quantum: the search for links, John A. Wheeler, *Proceedings of the Third International Symposium on the Foundations of Quantum Mechanics*, Tokyo, pp. 354–68 (1989), available at <https://philpapers.org/archive/WHEIPQ.pdf>

Molecular computation of solutions to combinatorial problems, L. M. Adleman, *Science* **266**, 1021 (1994)

Feynman Lectures in Computation, published in paperback by Westview Press (2000)

Negative differential resistance on the atomic scale: implications for atomic scale devices, In-Whan Lyo and Phaedon Avouris, *Science* **245**, 1369 (1989)

Molecule cascades, A. J. Heinrich, *Science* **298**, 1381 (2002)

A kilobyte rewritable atomic memory, F. E. Kalff et al., *Nature Nanotechnology* **11** 926 (2016)

Quantum holographic encoding in a two-dimensional electron gas, C. R. Moon et al., *Nature Nanotechnology* **4**, 167 (2009)

Binary atomic silicon logic, T. Huff et al., *Nature Electronics* **1**, 636 (2018)

Spin mapping at the nanoscale and atomic scale, Roland Wiesendanger, *Rev. Mod. Phys.* **81**, 1495 (2009)

Coherent spin manipulation of individual atoms on a surface, Kai Yang et al., *Science* **366**, 509 (2019)

Silicon quantum computing: <https://sqc.com.au>

Chapter 5: Nanomachines

How to build and race a fast nanocar, G. J. Simpson, *Nature Nanotechnology* **12**, 604 (2017)

Kinesin motor transports vesicle along microtubule, <https://www.youtube.com/watch?v=plvQCOE9s_k>

Supraheroes, Emma Stoye, *Chemistry World* 14 Oct. 2016, <https://www.chemistryworld.com/features/supraheroes/1017562.article>

Brownian ratchets: Darwin's motors, George Oster, *Nature* **417**, 25 (2002)

Construction of artificial cilia from microtubules and kinesins through a well-designed bottom-up approach, R. Sasaki et al., *Nanoscale* **10**, 6323 (2018)

Powering an Inorganic Nanodevice with a Biomolecular Motor, Ricky K. Soong et al., *Science* **290**, 1555 (2000)

Scientific Background on the Nobel Prize in Chemistry 2016, Royal Swedish Academy of Sciences, <https://www.nobelprize.org/uploads/2018/06/advanced-chemistryprize2016-1.pdf.>

Chemical topology: complex molecular knots, links, and entanglements, Ross S. Forgan, Jean-Pierre Sauvage, and J. Fraser Stoddart, *Chem. Reviews* **111**, 5434 (2011)

Light-driven monodirectional molecular rotor, N. Koumura, *Nature* **401**, 152 (1999)

Chapter 6: Are the nanobots nigh?

The debate between Smalley and Drexler was published in *Chemical & Engineering News* in Dec. 2003: <http://pubsapp.acs.org.ezproxy.nottingham.ac.uk/cen/coverstory/8148/8148counterpoint.html>

Nanosystems: Molecular Machinery, Manufacturing, and Computation, K. Eric Drexler (John Wiley & Sons, Inc., 1992)

Small problems, Philip Ball, *Nature* **362**, 123 (1993). A review of Drexler's *Nanosystems*.

Stereodivergent synthesis with a programmable molecular machine, S. Kassem et al., *Nature* **549**, 374 (2017)

Pick-up, transport and release of a molecular cargo using a small-molecule robotic arm, S. Kassem et al., *Nature Chemistry* **8**, 138 (2016)

Autonomous robotic nanofabrication with reinforcement learning, P. Leinen et al., *Science Advances* **6**, eabb6987 (2020)

Index

Index

SUPERCONDUCTIVITY
A Very Short Introduction
Stephen J. Blundell

Superconductivity is one of the most exciting areas of research in physics today. Outlining the history of its discovery, and the race to understand its many mysterious and counter-intuitive phenomena, this *Very Short Introduction* explains in accessible terms the theories that have been developed, and how they have influenced other areas of science, including the Higgs boson of particle physics and ideas about the early Universe. It is an engaging and informative account of a fascinating scientific detective story, and an intelligible insight into some deep and beautiful ideas of physics.

THE LAWS OF THERMODYNAMICS

A Very Short Introduction

Peter Atkins

From the sudden expansion of a cloud of gas or the cooling of a hot metal, to the unfolding of a thought in our minds and even the course of life itself, everything is governed by the four Laws of Thermodynamics. These laws specify the nature of 'energy' and 'temperature', and are soon revealed to reach out and define the arrow of time itself: why things change and why death must come. In this *Very Short Introduction* Peter Atkins explains the basis and deeper implications of each law, highlighting their relevance in everyday examples. Using the minimum of mathematics, he introduces concepts such as entropy, free energy, and to the brink and beyond of the absolute zero temperature. These are not merely abstract ideas: they govern our lives.

> 'It takes not only a great writer but a great scientist with a lifetime's experience to explains such a notoriously tricky area with absolute economy and precision, not to mention humour.'
>
> **Books of the Year, Observer.**

www.oup.com/vsi

RELATIVITY
A Very Short Introduction
Russell Stannard

100 years ago, Einstein's theory of relativity shattered the world of physics. Our comforting Newtonian ideas of space and time were replaced by bizarre and counterintuitive conclusions: if you move at high speed, time slows down, space squashes up and you get heavier; travel fast enough and you could weigh as much as a jumbo jet, be squashed thinner than a CD without feeling a thing - and live for ever. And that was just the Special Theory. With the General Theory came even stranger ideas of curved space-time, and changed our understanding of gravity and the cosmos. This authoritative and entertaining *Very Short Introduction* makes the theory of relativity accessible and understandable. Using very little mathematics, Russell Stannard explains the important concepts of relativity, from E=mc2 to black holes, and explores the theory's impact on science and on our understanding of the universe.

SCIENTIFIC REVOLUTION
A Very Short Introduction
Lawrence M. Principe

In this *Very Short Introduction* Lawrence M. Principe explores the exciting developments in the sciences of the stars (astronomy, astrology, and cosmology), the sciences of earth (geography, geology, hydraulics, pneumatics), the sciences of matter and motion (alchemy, chemistry, kinematics, physics), the sciences of life (medicine, anatomy, biology, zoology), and much more. The story is told from the perspective of the historical characters themselves, emphasizing their background, context, reasoning, and motivations, and dispelling well-worn myths about the history of science.

www.oup.com/vsi

NUCLEAR POWER
A Very Short Introduction
Maxwell Irvine

The term 'nuclear power' causes anxiety in many people and there is confusion concerning the nature and extent of the associated risks. Here, Maxwell Irvine presents a concise introduction to the development of nuclear physics leading up to the emergence of the nuclear power industry. He discusses the nature of nuclear energy and deals with various aspects of public concern, considering the risks of nuclear safety, the cost of its development, and waste disposal. Dispelling some of the widespread confusion about nuclear energy, Irvine considers the relevance of nuclear power, the potential of nuclear fusion, and encourages informed debate about its potential.

www.oup.com/vsi

Science and Religion

A Very Short Introduction

Thomas Dixon

The debate between science and religion is never out of the news: emotions run high, fuelled by polemical bestsellers and, at the other end of the spectrum, high-profile campaigns to teach 'Intelligent Design' in schools. Yet there is much more to the debate than the clash of these extremes. As Thomas Dixon shows in this balanced and thought-provoking introduction, many have seen harmony rather than conflict between faith and science. He explores not only the key philosophical questions that underlie the debate, but also the social, political, and ethical contexts that have made 'science and religion' such a fraught and interesting topic in the modern world, offering perspectives from non-Christian religions and examples from across the physical, biological, and social sciences.

'A rich introductory text . . . on the study of relations of science and religion.'

R. P. Whaite, Metascience

PLANETS
A Very Short Introduction
David A. Rothery

This *Very Short Introduction* looks deep into space and describes the worlds that make up our Solar System: terrestrial planets, giant planets, dwarf planets and various other objects such as satellites (moons), asteroids and Trans-Neptunian objects. It considers how our knowledge has advanced over the centuries, and how it has expanded at a growing rate in recent years. David A. Rothery gives an overview of the origin, nature, and evolution of our Solar System, including the controversial issues of what qualifies as a planet, and what conditions are required for a planetary body to be habitable by life. He looks at rocky planets and the Moon, giant planets and their satellites, and how the surfaces have been sculpted by geology, weather, and impacts.

"The writing style is exceptionally clear and pricise"

Astronomy Now

Nuclear Weapons
A Very Short Introduction
Joseph M. Siracusa

In this *Very Short Introduction*, the history and politics of the bomb are explained: from the technology of nuclear weapons, to the revolutionary implications of the H-bomb, and the politics of nuclear deterrence. The issues are set against a backdrop of the changing international landscape, from the early days of development, through the Cold War, to the present-day controversy of George W. Bush's National Missile Defence, and the threat and role of nuclear weapons in the so-called Age of Terror. Joseph M. Siracusa provides a comprehensive, accessible, and at times chilling overview of the most deadly weapon ever invented.

www.oup.com/vsi

INNOVATION
A Very Short Introduction
Mark Dodgson & David Gann

This *Very Short Introduction* looks at what innovation is and why it affects us so profoundly. It examines how it occurs, who stimulates it, how it is pursued, and what its outcomes are, both positive and negative. Innovation is hugely challenging and failure is common, yet it is essential to our social and economic progress. Mark Dodgson and David Gann consider the extent to which our understanding of innovation developed over the past century and how it might be used to interpret the global economy we all face in the future.

'Innovation has always been fundamental to leadership, be it in the public or private arena. This insightful book teaches lessons from the successes of the past, and spotlights the challenges and the opportunities for innovation as we move from the industrial age to the knowledge economy.'

Sanford, Senior Vice President, IBM

GALAXIES
A Very Short Introduction
John Gribbin

Galaxies are the building blocks of the Universe: standing like islands in space, each is made up of many hundreds of millions of stars in which the chemical elements are made, around which planets form, and where on at least one of those planets intelligent life has emerged. In this *Very Short Introduction*, renowned science writer John Gribbin describes the extraordinary things that astronomers are learning about galaxies, and explains how this can shed light on the origins and structure of the Universe.

FORENSIC SCIENCE
A Very Short Introduction
Jim Fraser

In this Very Short Introduction, Jim Fraser introduces the concept of forensic science and explains how it is used in the investigation of crime. He begins at the crime scene itself, explaining the principles and processes of crime scene management. He explores how forensic scientists work; from the reconstruction of events to laboratory examinations. He considers the techniques they use, such as fingerprinting, and goes on to highlight the immense impact DNA profiling has had. Providing examples from forensic science cases in the UK, US, and other countries, he considers the techniques and challenges faced around the world.

An admirable alternative to the 'CSI' science fiction juggernaut...Fascinating.

William Darragh, Fortean Times

www.oup.com/vsi